THE STATE AND ECONOMIC LIFE

EDITORS: Mel Watkins, University of Toronto; Leo Panitch, Carleton University

5 LAUREL SEFTON MACDOWELL

'Remember Kirkland Lake'
The Gold Miners' Strike of 1941–42

On 18 November 1941, the gold miners of Kirkland Lake struck for union recognition. The Kirkland Lake strike was a bitter struggle between the mine operators and their employees and became a national confrontation between the federal government and the labour movement over the issue of collective bargaining.

Locally, the dispute was affected by the company-town environment and by the mine operators' paternalistic view of labour relations. Through the difficult winter months, the community – polarized by the events – tried to deal with both the 'political' and social impact of the conflict. The author's father, Larry Sefton, emerged as one of the local leaders of the strike, which itself was a training ground for many future trade unionists.

The strike was waged in the special circumstances of the war economy, and was a microcosm of wartime developments, which produced unprecedented union growth, serious industrial unrest, hostile management response, and generally antagonistic labour/government relations.

Professor MacDowell shows that, even though the strike was lost, its eventual effect on labour policy gave the dispute its particular significance. To win the strike, government intervention and the introduction of collective bargaining were necessary, yet the only intervention was by the Ontario Provincial Police, who were ordered to assist the mining companies to operate with strike-breakers. The federal government refused to intervene, in spite of virtually unanimous support for the strike by the Canadian labour movement.

MacDowell concludes that the strike succeeded in unifying organized labour behind the demand for collective-bargaining legislation. It highlighted the inadequacy of the government's wartime labour policy, and ultimately forced the government to authorize collective bargaining, first for Crown companies and then for all industrial workers. Thus, the Kirkland Lake strike was not only an important wartime dispute affecting policy development, but it also established a special legacy for trade unionists as part of the history of their movement.

LAUREL SEFTON MACDOWELL is Assistant Professor of History at the University of Toronto.

THE STATE AND ECONOMIC LIFE

Editors: Mel Watkins, University of Toronto; Leo Panitch, Carleton University

This series, begun in 1978, includes original studies in the general area of Canadian political economy and economic history, with particular emphasis on the part played by the government in shaping the economy. Collections of shorter studies, as well as theoretical or internationally comparative works, may also be included.

LAUREL SEFTON MACDOWELL

'Remember Kirkland Lake': the History and Effects of the Kirkland Lake Gold Miners' Strike, 1941–42

UNIVERSITY OF TORONTO PRESS

Toronto Buffalo London

© University of Toronto Press 1983
Toronto Buffalo London
Printed in Canada

ISBN 0-8020-5585-0 (cloth)
ISBN 0-8020-6457-4 (paper)

Canadian Cataloguing in Publication Data

MacDowell, Laurel Sefton, 1947–
'Remember Kirkland Lake'
Bibliography: p.
Includes index.
ISBN 0-8020-5585-0 (bound). – ISBN 0-8020-6457-4 (pbk.)
1. Kirkland Lake (Ont.) – Gold Miners' Strike, 1941–1942.
2. Collective bargaining – Mining industry – Canada –
History – 20th century. I. Title.
HD5329.M62 1941 K57   331.89′28223422′09713144   C82-094902-7

Contents

Preface

This monograph is a history of the Kirkland Lake gold miners' strike of 1941–42, which at the time attracted the sympathy and support of the entire labour movement – particularly the new industrial unions of the Canadian Congress of Labour (CCL). This strike provided a focus for the labour movement's discontent, contributed to its development of common legislative objectives, and eventually influenced Canada's wartime and post-war labour policy. This policy established a new legislative framework for collective bargaining that has been modified only slightly since that time.

In reality the strike was two strikes: one at the community level and one at the national level. The problems created for the gold miners' union by inadequate industrial relations legislation so affected the local situation that the strike became more than just an interesting conflict in a fairly typical mining community. It became a milestone in the struggle for legislation to protect and promote free collective bargaining. As H.A. Logan has suggested, 'Kirkland Lake marked the low point in industrial relations for the war. But from it began the march toward P.C. 1003.'[1] The proclamation of PC 1003 (this war order was the first federal legislation recognizing and supporting the practice of collective bargaining) cannot be attributed solely to the loss of the Kirkland Lake strike. There were many other pressures operating on the government in 1943–44 – not the least of which was the growing influence of the CCF (Co-operative Commonwealth Federation). However, the strike did unify the labour movement and crystallize its discontent with the existing legislative framework. It prompted the government to proclaim PC 10802, which extended the right of collective bargaining to employees in Crown corporations. Once this precedent had been set, PC 1003 became inevitable.[2] Consequently, the events of the strike provide insight into the evolution of the nation's wartime labour policy.

In order to appreciate the evolution of this policy, it is insufficient to consider simply the political debate or the crises that precipitated the change. Even the important strikes, which crystallized labour's discontent and prompted specific concessions, took place within the special context of the war economy and a general realignment of industrial and political forces. Economic tensions associated with the war generated pressures for reform that could not be contained.

In many respects, the Kirkland Lake strike was a wartime phenomenon, for the war years were a period of antagonistic labour/government relations and serious industrial unrest, which labour attributed to wage controls, the failure of the government to consult labour on policies directly affecting employees, and the inadequacy of the existing collective-bargaining legislation. At the centre of this conflict, as in Kirkland Lake, was the demand for collective bargaining, which was not merely a means of raising wages and improving working conditions. It was a demand by organized workers for a new status and the right to participate in decision making, both in industry and in government. It became an issue not only on the shop floor, where employers and unions met directly, but also in the political arena. In 1941, the Kirkland Lake miners were part of a growing industrial union movement, responding to the pressures of the war economy and to existing government policy. They were aware of the American legislation supporting collective bargaining, but did not initially regard their dispute as a 'watershed' or a struggle for a change in government policy. Nevertheless, this is what the strike became.

This study is more detailed than was originally intended, for several reasons. Little has been written about the strike elsewhere, and I discovered a number of valuable and virtually unused records made at the time by informed observers who considered the strike a significant event. These sources – which included the J.L. Cohen papers and the J.R. Mutchmor papers – confirmed my own view that a more thorough inquiry would be worth while. My intention was to consider not only the union's organizational activity and the inadequacy of existing legislation, but also the activities, views, and feelings of the individuals involved. This required the inclusion of personal and anecdotal material that might have been eliminated in an analytical presentation, but has been included here to give a better sense of what this strike involved for the participants.

I have consciously emphasized the organizational theme – how the union was organized and the obstacles it faced – because these concerns were preeminent to trade unionists at that time. I have alluded to labour's political involvement with the CCF and the communists where it was relevant to the strike situation. However, such subjects are treated comprehensively in other works and are not the focal point of this study. Also, while the International Union of Mine Mill and Smelter Workers (IUMMSW) was a political union, weakened internally by conflicts between communists and socialists both before and after the strike, during the period covered in this study, the local's political factions agreed to minimize their differences in the interest of organizing a viable union and winning a master agreement. I have endeavoured to apply industrial relations theory and materials to this historical study where I thought that such an approach would illuminate and clarify the historical events.

The organization of this book is self-explanatory. For the most part it is chronological. There is some overlapping of information, but where that has occurred the facts are used to discuss different aspects of the strike.

I am indebted to the University Research Committee of the Canada Department of Labour for its financial assistance during the preparation of this study. This book has been published with the help of a grant from the Social Science Federation of Canada, using funds provided by the Social Sciences and Humanities Research Council of Canada, and a grant from the Publications Fund of the University of Toronto Press.

I owe much appreciation to many people for their assistance in conducting this research. I particularly want to thank Nancy Stunden and Danny Moore for their invaluable aid at the Public Archives of Canada, Bob and Kay Carlin for a delightful day in Gowganda with much good talk, and Professor Kenneth McNaught for his encouragement and advice. Thanks are owing to my mother, Elaine Sefton, for her typing efforts and her informed criticism, and to Anne McMaster for her expert typing of the final draft. Most particularly, I owe a great deal to my husband, Rick MacDowell, for his encouragement, his constructive criticism and enthusiasm, and more practically for his hours of babysitting.

The inspiration and motivation for this study were partly personal. The Kirkland Lake strike was the event that propelled my late father, Larry Sefton (at that time a twenty-four-year-old miner), into his thirty-year career in the labour movement. It is to his memory that this work is dedicated.

L.S.M.

Abbreviations

ACCL	All Canadian Congress of Labour
ACWA	Amalgamated Clothing Workers of America
AFL	American Federation of Labor
CBRE	Canadian Brotherhood of Railway Employees
CCF	Co-operative Commonwealth Federation
CCL	Canadian Congress of Labour
CIO	Congress of Industrial Organizations
CMA	Canadian Manufacturers' Association
FCSO	Fellowship for a Christian Social Order
IDI Act	Industrial Disputes Investigation Act
IDIC	Industrial Disputes Inquiry Commission
ILGWU	International Ladies' Garment Workers' Union
IUMMSW	International Union of Mine, Mill and Smelter Workers (Mine Mill)
IWW	Industrial Workers of the World
MWUC	Mine Workers' Union of Canada
NASCO	National Steel Car Corporation
NIRA	National Industrial Recovery Act (American)
NLB	National Labor Board (American)
NLRB	National Labor Relations Board (American)
NLSC	National Labour Supply Council
NSS	National Selective Service
NWLB	National War Labour Board (Canadian)
PWOC	Packinghouse Workers' Organizing Committee
SWOC	Steelworkers' Organizing Committee
TLC	Trades and Labour Congress
UAW	United Automobile Workers

UBRE	United Brotherhood of Railway Employees
UEW	United Electrical Workers
ULFTA	Ukrainian Labour Farm Temple Association
UMWA	United Mine Workers of America
WFM	Western Federation of Miners
WPTB	Wartime Prices and Trade Board
WUL	Workers' Unity League

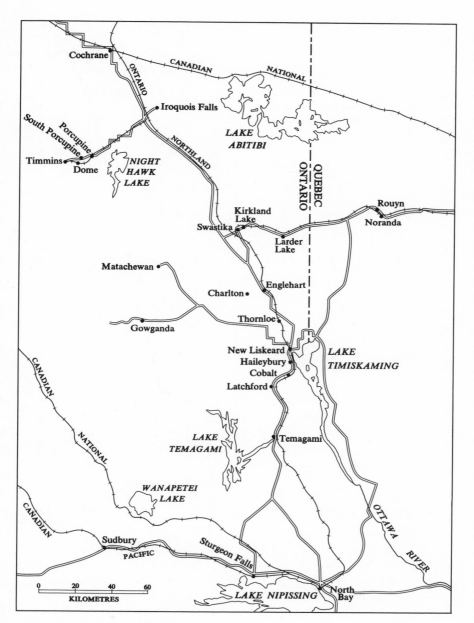

Northeastern Ontario, c. 1942

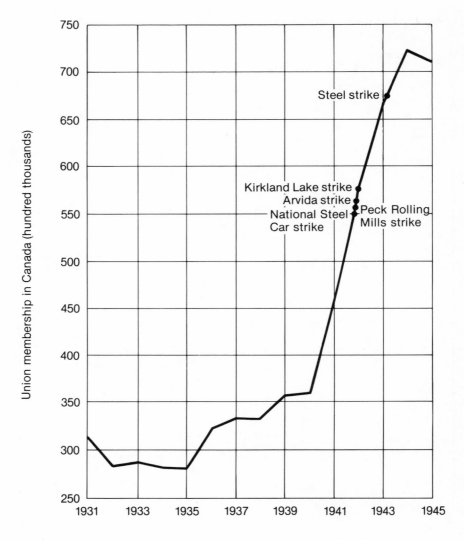

Union membership, 1931–45. This and the following graphs are, in one sense, a portrait of labour relations during the war. Just as the level of union membership reached a peak in 1944, the level of industrial conflict and the extent of labour's alienation from the government peaked only slightly earlier, in late 1943. The Kirkland Lake strike contributed substantially to the 1943 confrontation. Source: Canada, Department of Labour, *Labour Organizations in Canada* (1974–75)

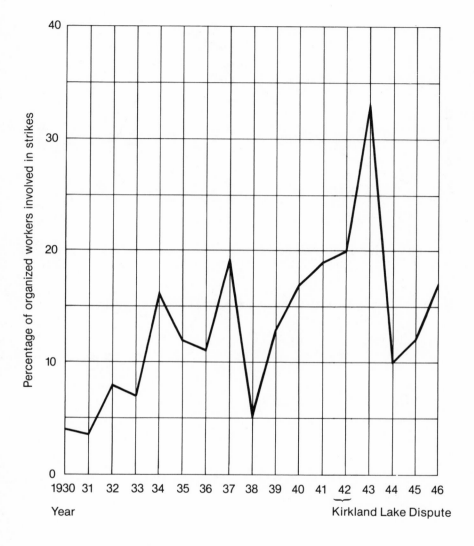

Industrial militancy/propensity to strike. Proportion of organized labour force engaging in strikes, 1930–46 (number of workers involved, as a percentage of total union membership). Source: Canada, Department of Labour, *Labour Organizations in Canada* (for union membership) and *Strikes and Lockouts in Canada* (for strikes)

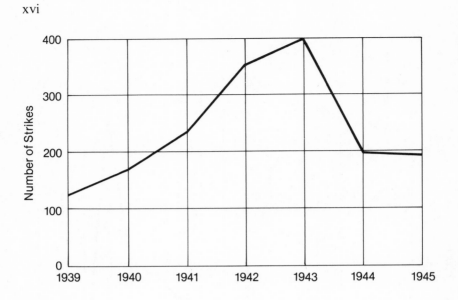

Number of strikes, 1939–45. Source: Canada, Department of Labour, *Strikes and Lockouts in Canada*

'REMEMBER KIRKLAND LAKE'

1
The wider context

One might as well tell the full grown man to resolve himself into a boy again and 'be seen and not heard' as tell labour it cannot have a voice in the management of industry through collective bargaining. Grass will grow, the river will reach the sea, the boy will become a man and labour will come into its own.[1]

We must have democracy in industry.[2]

My observation leads me to believe that while there are many contributing causes to (industrial) unrest, there is one cause which is fundamental. That is the necessary conflict – the contrast – between our political liberty and our industrial absolutism.[3]

A dispute of the magnitude of the Kirkland Lake strike is always a significant industrial relations event. The strike lasted almost three months, involved close to 4000 workers,[4] and resulted in a loss of 136 000 man-days – 45 per cent of the total time lost from all strikes in Ontario during the fiscal year of 1942.[5] The strike occurred at a time when gold mining was still classified as an 'essential' war industry, and the defeat of Local 240 of the International Union of Mine Mill and Smelter Workers (Mine Mill) undoubtedly hampered the union's attempts to organize in Northern Ontario. But the strike had an impact beyond the local community. Its importance can only be understood by considering several elements: the Canadian trade union and industrial relations setting at the beginning of the war; the American scene with its legislative precedent, the National Labor Relations Act (the Wagner Act); the pressures of the war economy and the resulting 'labour problem.' Each of these essential aspects of the industrial relations environment influenced the course of the Kirkland Lake strike and provided the context in which the events of the strike itself must be examined.

THE CANADIAN INDUSTRIAL RELATIONS SETTING

The political and economic climate in Canada in the 1930s was not conducive to trade union organization or the emergence of an innovative industrial union movement. Not only was there sluggish union growth and no comprehensive labour policy, but there was also no Roosevelt and no New Deal. The introduction of the Canadian welfare state was haphazard, took a long time to be implemented, and consequently provided little relief for workers. R.B. Bennett's proposals, modelled on the American experience and including minimum-wage legislation, were declared unconstitutional and Mackenzie King did not introduce the 'baby bonus' until 1944 (a period of relative prosperity).[6]

In the early 1930s, the trade union movement went on the defensive in order to survive.[7] Depression and unemployment resulted in a decline in union membership. There was a slight recovery in 1936 and membership peaked in 1937 – the year of the significant Oshawa strike – by which time the great American strikes by the new industrial CIO (Congress of Industrial Organizations) unions were over.[8] However, renewed recession in 1939 again adversely affected growth. At the outbreak of the war there were only 359000 organized workers, barely 17 per cent of the non-agricultural labour force. Whereas labour organization in the United States in the 1930s, accompanied by positive collective bargaining legislation, helped American workers to become more fully organized than Canadian workers, the Canadian labour movement would not catch up until the war production effort expanded the labour force and provided a favourable organizing climate.

It was the war that virtually eliminated unemployment so that unions would grow at an accelerated rate. Trade union organization during the war resulted from the expansion of new industrial unions associated with the American CIO. In 1937, CIO organizers came to Canada to assist auto workers in their attempt to organize.[9] This led to the organization of many other semi-skilled industrial employees. The prime minister considered the new union leaders to be irresponsible; the press and business believed they were opportunistic agitators. Nevertheless, between 1939 and 1946, union membership nearly doubled to 832000 organized workers. A further increase occurred between 1945 and 1949[10] so that by 1949 the number of union members was almost three times what it had been in 1937.

The greatest membership increases would occur in 1941 and 1942 at the time of the Kirkland Lake strike. By 1941, there were 461 681 union members in Canada, an increase of 96 137 (27.5 per cent) over 1940. This represented a rate of increase three times greater than that of 1940 over 1939. In 1941, unionization in the mining industry increased 8.4 per cent.

These new industrial unions expanded with the industries from which they drew their support.[11] In 1942, union membership increased by more than 25 per cent over 1941. As in the previous year, unionization continued to be greatest in the metals industries (especially in steel fabricating and wartime aircraft plants), where membership increased by 99 per cent. The increase in mining organization was less significant in 1942. Union membership actually dropped from 38 678 members (or 8.4 per cent of all union members) in 1941 to 34 915 members (or 6 per cent) in 1942. This decline was due, in part, to the failure of Mine Mill organizing drives in Kirkland Lake and elsewhere. The other mining union, the United Mine Workers (UMWA), generally maintained its strength. In 1942 the demand for strategic materials led to a greater concentration on the mining of basic metals and the classification of the precious metals mining as 'non-essential' industries. These mines (which included the Kirkland Lake group) cut back their operations and the number of men employed was reduced drastically. Union membership in the mining industry remained fairly constant during 1943–45 at between 36 000 and 38 000 members, while at the same time membership in manufacturing soared. By 1942 the steelworkers, auto workers, and machinists had formed the largest unions in Canada[12] representing three-fifths of the total union membership in the metals industries. By 1946, these were the most highly unionized industries, a growth of union membership directly related to war production.[13]

The increased organizational activity met with considerable employer resistance and resulted in unprecedented levels of industrial conflict. Until the government passed legislation supporting collective bargaining in 1944, there was a continuous increase in the number of strikes, workers involved, and man-days lost. The level of industrial unrest climbed in 1941 and 1942 and reached a new peak in 1943.[14] In 1940 there were 168 strikes involving 60 619 workers and 266 318 man-days lost. In 1941 there were 231 strikes involving 87 091 workers and 433 914 man-days lost. In 1942 the number of strikes increased to 354 involving 133 916 workers with a loss of 450 202 man-days.

In 1941 the strikes occurred primarily in coal and gold mining, metal fabricating, woodworking, and the food industries. The Kirkland Lake strike was one of the ten big strikes of that year, and accounted for 78 000 lost man-days. The Kirkland Lake strike and the Ford strike in Windsor accounted for almost one-third of all man-days lost.[15] However, in 1941 mining disputes accounted for 44 per cent (191 689 man-days) of all man-days lost, while in 1942 such disputes accounted for only 28 per cent (129 529 man-days) of the total time lost. Despite the increase in industrial conflict in 1942, industrial production rose 28 per cent and employment rose 14 per cent.

In addition to man-days lost, it is useful to consider the number of employees involved in strikes as a proportion of the total union membership. This figure is a rough indicator of industrial militancy or propensity to strike, and indicates whether discontent is generalized throughout the labour movement. In 1938, approximately 5 per cent of organized employees engaged in strike activity (i.e. one in twenty), while in 1939, 13 per cent did so (i.e. one in eight). In 1941 and 1942 close to 20 per cent of the (now expanded) membership engaged in strikes (i.e. one in five). In 1943, the year before PC 1003 (see appendix 1) was proclaimed, one out of every three trade union members was involved in strike activity – a level of membership involvement exceeded only in 1919 and then only marginally. To the extent that membership involvement in industrial conflict is a measure of employee disaffection, 1919 is the only year with which 1943 can be compared.[16]

Government statistics concerning the cause of a strike can be misleading, since an employer refusal to accede to a wage demand may well be a refusal to recognize a new union or enter into a first collective agreement. Nevertheless, the frequency with which 'recognition' is mentioned or singled out as an independent issue is significant. In 1941, wages were listed as the major 'cause' of 113 strikes involving 30 362 workers and a time loss of 147 450 man-days. Union recognition caused the next highest number of strikes in that year. This represented nearly twice as many strikes for that reason as in 1940 and an increase in time lost of over 25 per cent.[17] It is also significant that most of these 'recognition' strikes eventually resulted in agreements with committees of employees. In 1942, there were even more recognition strikes: 174 disputes over wages and 43 recognition strikes (which resulted in a time loss of 112 538 man-days). Of these, 6 were successful, *18* (including that of the Kirkland Lake miners) were unsuccessful, and 14 resulted in 'compromise.'[18]

The Kirkland Lake strike was, in many respects, typical of what was happening on the industrial scene throughout Canada. It was for this reason that it evoked the sympathy of trade unionists across the country, and became the focus for a national debate on the inadequacy of the federal government's labour policy.

Such substantial union growth transformed the policies and *structure* of the Canadian labour movement in the process. These structural changes had both organizational and political ramifications that affected the course of the Kirkland Lake dispute. Many newly organized employees became affiliated as industrial unions to the craft-oriented Trades and Labour Congress (TLC). By 1939 there were 22 000 members in the TLC who belonged to CIO international unions.[19] The fundamental difference in outlook between craft and industrial unions made inter-union rivalry inevitable.

The power of craft unions was based on their control of the supply of skilled tradesmen. Since their skills could not be easily acquired, or easily replaced, these tradesmen, acting together, had considerable bargaining power. This system of job control depended upon the definition and maintenance of an exclusive jurisdiction for each craft and union, and this function was performed by the prevailing central labour body (the AFL [American Federation of Labor] in the United States and the TLC in Canada). Industrial unionism implied a new system. Industrial unionists sought to organize the mass production industries on an industrial basis and allow the workers in the plants to elect their bargaining representatives on the basis of majority rule. Their bargaining power depended not upon a monopoly of skills, but rather a control of numbers. Although small numbers of unskilled workers could be easily replaced, an employer would not usually be in a position to acquire an entire new work force.

In theory and in fact, exclusive craft jurisdiction was in conflict with the election of bargaining representatives on the basis of majority rule.[20] Under the old system, the AFL–TLC determined which union would be chartered as the workers' representative, and even then it would only represent those who worked at or in association with the craft. Consequently some workers could be organized only as a 'tag end' to a particular craft, and many could not be organized at all. Unskilled workers, who made up the vast majority of employees in the mass production industries, saw no reason why their effort to organize should be frustrated because of the concerns of a minority of skilled workers.

Such different union perspectives made the craft unions in the TLC cautious about the legislative changes proposed by the new unions. Craft unions were unwilling to support majority-rule principles that would submerge skilled employees in a sea of unskilled production workers. Nevertheless, the industrial unions pressured the TLC to support legislation patterned after the American Wagner Act. In 1937, the TLC responded to the pressure of its industrial unions be drafting a 'Model Bill' and presenting it to various provincial governments. This bill was important firstly as a legislative recommendation and secondly as an issue that accentuated conflict within the TLC between the craft and industrial unions.

There were serious constitutional barriers to the creation of a centralized, industrial relations system – an impediment that the King government of the 1930s used as a convenient excuse to forestall any legislative action on collective bargaining.[21] The presentation of the TLC 'Model Bill' was ignored by the federal government but resulted in a spate of new provincial labour acts passed between 1937 and 1939, and a Criminal Code amendment. Provincial legislation provided for *some* recognition for trade unions and collective bargaining, but such laws were largely unenforced and therefore ineffective. The Canadian trade union movement 'had attempted to employ the legislative arm to foster and protect trade union associations. Experience born of failures during these years indicated the patterns that had to be employed if legislation was to be an effective instrument to assist in trade union organization.'[22] Section 502A of the Criminal Code (enacted in 1939 as a result of pressure from the CCF)[23] made it an indictable offence for an employer to refuse to employ or dismiss or intimidate any person for 'the sole reason' that he was a member of a union; however the criminal burden of proof made it virtually impossible to secure a conviction. Even if an employer were found guilty, he could only be *penalized*. There was no *remedy* (such as reinstatement) for the employee.[24] This provision was ineffective in protecting a worker's right to join a union.[25] Despite the North American environment, and American industrial relations developments, at the outbreak of the War Canadian statesmen still supported the British system of 'voluntarism.'

The 1937 'Model Bill' was sharply criticized at two successive TLC conventions[26] because it did not *compel* employers to bargain with unions with majority support, or expressly prohibit 'company unions'; nor did it include machinery to determine the bargaining agent when jurisdictional disputes arose between unions. The new unions, which lacked the economic strength to establish collective bargaining relationships, required active government intervention to protect their organizations. The craft unions, which were strong and entrenched, did not need government intervention to gain recognition from employers, and were wary of the increased role of government implicit in the 'Wagner' principles. President Draper cautioned that features of the Wagner Act would not be workable in Canada. He was concerned about keeping the goodwill of major international craft unions on whom the TLC was financially dependent. These unions were affiliated with the AFL in the United States, and took the position that the determination of bargaining units on other than a craft basis should not be the concern of government. He was concerned (correctly as it turned out) that if a government agency defined the group of employees appropriate for collective bargaining, and among whom the union had to demonstrate majority support, there would be little attention given to the unique interests of skilled employees, and the voting constituency might be so drawn that no union could secure majority support. The voting constituency issue arose in Kirkland Lake.

In 1939 the conflict between craft and industrial forms of organization resulted in the expulsion from the TLC of the international industrial unions affiliated with the CIO.[27] The unions expelled included the UMWA, the Amalgamated Clothing Workers, Mine Mill, and the Steelworkers' Organizing Committee. Thus, the great split in the American labour movement was duplicated in Canada.[28] Subsequently, the craft unions also supported legislative protection for collective bargaining, but not until they had embarked on a more dynamic organizing policy and had begun to broaden their own organizational base.[29] The unity of the TLC and the CCL (Canadian Congress of Labour) on this legislative issue was achieved by the end of the Kirkland Lake strike. The loss of this strike concerned the affiliates of both congresses, for they saw it as a threat to all union-organizing activities.

In 1939 the membership of all the industrial unions outside the TLC amounted to approximately 55 000 in sixteen different unions.[30] Thus, at its second convention, the CIO decided to authorize the establishment of a formal CIO committee in Canada. On organizational matters the committee was to be funded and influenced by the CIO. On legislative matters it was to have complete autonomy. The Canadian CIO committee was formed on 5 November 1939 at a conference held in Ottawa and chaired by Canadian trade unionist Silby Barrett (the director of the Maritimes district of the United Mineworkers of America [UMWA]). The conference was attended by 105 CIO union representatives working primarily in eastern Canada. Barrett became the head of the Canadian 'CIO.' On this more formal basis, industrial unionism in Canada continued to grow.

The CIO committee began almost immediately to seek 'closer co-operation'[31] with the All Canadian Congress of Labour (ACCL). This small labour congress was founded in 1927 to promote the interests of independent national unions. By 1928 the ACCL had accepted industrial organization as a basic principle, and its affiliates were mostly industrial unions; but the ACCL had been weakened by internal disputes. In 1936 a breakaway group established the ultra-national, and conservative Canadian Federation of Labour.[32] Organizationally, the ACCL was at a standstill, but it wanted to take advantage of the war situation to expand its membership.

The outcome of the discussions between the Canadian CIO and the ACCL was the formation of a provisional committee of six members – three from each organization – to complete negotiations for the fusion of the two bodies and to draft a new constitution.[33] At its founding convention in September 1940, the constitution of the Canadian Congress of Labour was approved. The unions that formed the CCL included the CBRE (Canadian Brotherhood of Railway Employees), the National Union of Operating Engineers, the Canadian Association of Railwaymen, and the Canadian Electrical Union from the ACCL. The CIO unions included the American Newspaper Guild, the UMWA, the Amalgamated Clothing Workers, the Steelworkers' Organizing Committee (SWOC), the United Auto Workers (UAW), the Rubber Workers' Organizing Committee, the United Electrical Workers (UEW), the Packinghouse Workers Organizing Committee (PWOC), and 'Mine Mill.' All of these CIO unions had emerged in the 1930s. The organizing committees soon became established industrial unions. The leaders of these committees became trade union officers and important allies when the Kirkland Lake miners sought to organize.

The CCL was determined 'to organize the unorganized' into established industrial unions, to promote the interests of the affiliated unions, and generally to advance the social and economic condition of workers in Canada. The new congress adhered to the joint principles of industrial unionism and legislative autonomy. This latter principle was insisted upon by the ACCL and continued its nationalistic tradition.

In pursuit of its organizational goal, the CCL in 1940[34] established a special organizing fund to assist unions seeking to organize new members. Among the groups granted money from this fund were the Kirkland Lake miners. The congress also organized *special* appeals to raise funds to support organizing drives and strikes. Its most ambitious appeal in these years involved the support of the Kirkland Lake strike. Sometimes congress organizers helped member unions to organize new locals, or affiliates organized directly chartered 'Congress local unions.' This relationship between the congress and its affiliates was reciprocal. Unlike the TLC, the CCL (with the acquiescence of its executive board, composed of the senior officers of member unions) became actively involved in organizing, as well as engaging in the more traditional lobbying activities of a central labour body.

The CCL executive constantly, though never quite successfully, tried to maintain an overview of the organizational activities of its affiliates and its own organizing staff.[35] However, events were moving too quickly. Congress officers were overworked and did not have the time to formulate an overall organizational strategy. Workers approached the CCL or one of its member unions for help in forming a local, and such situations would have to be met at once. Nevertheless, it was clear that under the spur of wartime full employment, much of the mining, basic steel, automotive, electrical, rubber, and parts of the clothing industries was becoming unionized[36] by CCL affiliates.[37]

The CCL grew quickly and became a permanent challenge to the TLC. In 1941, the TLC reported that it had 144 592 members.[38] The CCL reported a membership of 125 000 although the labour department calculated only 97 243 members. This latter figure still represented a gain of 20 260 members over 1940. In 1942 the CCL's membership was still in second place, but it was a large, rapidly growing organization composed of 6 national and 13 international unions, 423 locals, and 141 directly chartered locals with a reported membership of 160 000. This figure represented an increase of 35 000 members over 1941. The labour department calculated that CCL membership was even larger, and indicated that the CCL consisted of 564 locals and 200 089 members, 161 999 workers in the affiliated unions, and 38 090 in directly chartered CCL locals.[39] Because of their industrial form of organization, affiliates of the CCL were now the largest unions in Canada.[40] The CCL's growth was much faster than the TLC because it had a heavy concentration of membership in the expanding mass-production industries, and was more aggressive in seeking new members.[41] The organizing campaigns in Kirkland Lake were very much a part of this boom in trade union organization, and should be understood in that context.

THE AMERICAN EXPERIENCE

One cannot fully appreciate Canadian labour relations developments during World War II without an understanding of events in the United States during the 1930s. Many of the issues and problems that arose in Canada had already been resolved in the United States. Canadian trade unions and employers were well aware of American events. Both parties developed their strategy against the background of the United States experience. Trade unions consistently referred to American precedents as a model for the resolution of similar problems in Canada. Employers, on the other hand, were equally adamant to preserve the status quo and avoid the adoption of American legislative solutions.

At the outset of the New Deal, neither Franklin Roosevelt nor Secretary of Labor Frances Perkins had expressed any particular interest in protecting labour's right to organize and bargain collectively. Their concern was initially limited to the enactment of social welfare legislation, and direct unemployment relief. However, as trade union organization increased (from 2.8 million members in 1933 to 8.4 million in 1941)[42] it became necessary to develop an increasingly sophisticated collective bargaining policy. In 1933, section 7a of the National Industrial Recovery Act (NIRA) proclaimed the right of employees to organize and bargain collectively through representatives of their own choosing. This provision was strongly opposed by employers who (with the assistance of the National Association of Manufacturers) encouraged the development of 'company unions' to offset the growth of bona fide independent trade unions, along the same lines as Mackenzie King's 'employee representation plan,' which was implemented during World War I at the Rockefeller Colorado Fuel and Iron Company. The 'Colorado Industrial Plan' was the 'most imaginative and best publicized of company union plans' and impeded independent unionism for eighteen years. As a result of such employer activities there was a great expansion of 'company unionism,'[43] which initially was successful in frustrating attempts to organize trade unions that were independent of employer influence.

By 1934 unions were actively organizing the unorganized workers in the steel, auto, trucking, textiles, and longshoring industries. But the legal effect of Section 7a remained uncertain, and labour's position remained insecure. The American Federation of Labor (AFL) was unwilling to organize large groups of workers who were clamouring for organization. The craft basis for organization was inadequate and obviously inappropriate for the purpose of organizing industrial workers.[44] The hybrid craft/industrial experiments known as federal unions failed, but the AFL refused to adopt John L. Lewis' proposal of industrial unionism. These disagreements between Lewis and the AFL executive eventually led to the formation of the Committee for Industrial Organization (CIO).

With the formation of the CIO in 1935 (initially within the AFL and later as a rival organization, in 1938), organized labour embarked on the enormous task of organizing the mass-production industries. In 1936–37 the CIO enjoyed phenomenal success. The Steel Workers' Organizing Committee (SWOC) organized most of the steel industry – the decisive achievement coming in the peaceful agreement with U.S. Steel on 21 March 1937. In 1937, the United Auto Workers (UAW), following the sit-down strike at Flint, entered into agreements with General Motors and Chrysler, as well as with many smaller firms. Similar breakthroughs were made in the textile, electrical, rubber, and woodworking industries.[45]

By 1937 the CIO posed a challenge to the dominance of the AFL. In response, the older federation initiated an organizing campaign of its own, which was also very successful.[46] Thus competition between the two rival federations contributed to union growth. The dramatic burst of trade union organization and increased membership was very similar to that which occurred in Canada several years later. As in Canada during the World War II, there were significant changes in the policy and structure of the American labour movement. Industrial unionism was accepted after several years of dissension and disunity within the AFL because it was demonstrated that it was viable. Some AFL affiliates (like the International Brotherhood of Electrical Workers and International Association of Machinists) abandoned pure craft unionism and adopted an industrial form of organization.

The rapid development of unions in the mass-production industries resulted in important changes on the shop floor. Organized workers gained significant wage increases, and bargaining narrowed differentials in earnings between unskilled, semi-skilled, and skilled workers. The seniority principle gave workers with long service some job security. It restricted management's previously unfettered discretion to hire, fire, promote, demote, or lay off employees. For the first time employers were called upon to justify their actions. Thus the most significant accomplishment of the new unions was the establishment of formalized grievance procedures:

At the outset this took the form of creating a shop steward system in the plant and of compelling the employer to deal with it in the disposition of grievances. This led shortly to the erection of hierarchically arranged steps with increased levels of authority on each side through which grievances passed in accordance with time limits. Towards the end of the period a growing but smaller number of collective bargaining agreements provided for arbitration as the terminal step in the procedure, utilizing an impartial person to render a final and binding award.[47]

The growth of industrial unionism and the innovations it spawned resulted in a fundamental alteration of the economic and social position of the industrial worker.

The new unionism resulted in increased political activity by the labour movement, which undermined the historic political neutrality of the AFL. John L. Lewis, the dominant labour leader in the mid and late 1930s, committed the CIO to an active political policy. President Franklin Roosevelt consciously made the urban working class a cornerstone of his New Deal coalition and encouraged their political activity. The close relationship between organized labour and the national government resulted in some labour appointments to government agencies that were given the tasks of trying to effect the country's recovery from the Depression and later of planning and implementing wartime policies in the United States.[48] Partnership in the coalition meant participation in the formation and administration of its policies, and throughout the 1930s, the Democrats developed legislation favourable to collective bargaining. There was no similar alliance between the CCL unions and the Liberal Party of Canada.[49]

The changes in American labour law in the late 1930s represented a fundamental shift in public policy. Prior to 1935 and the passage of the Wagner Act, labour boards built up a body of 'labour common law' which later was embodied and elaborated in the Wagner Act. Among the first issues dealt with by these agencies was the question of the appropriate bargaining unit, or unit of employees who shared a community of interest and should, therefore, bargain together. Closely associated was the question of whether there should be several unions in a particular plant, or a single bargaining agent. Some administrators advocated proportional representation so that a union that received a majority of votes would represent the majority, while minorities and individuals retained the right to bargain on their own behalf. The labour boards adopted the concept of an *exclusive* bargaining agency for the union that enjoyed the majority of the employees in the bargaining unit, because it eliminated the fragmentation, internal rivalries, and conflict implicit in multi-unionism, and ensured more orderly collective bargaining. They affirmed the right of workers to collective bargaining and concluded that employers had a positive duty to recognize and to bargain with their workers' selected representatives. These decisions had little immediate value since they could not be *enforced* against a determined and resourceful employer, and remained a mere declaration of rights. These early boards had responsibility without authority. The board procedures themselves were a problem and caused delays that gave an advantage to the anti-union employer 'since a discharged employee or a new unstable union required prompt remedies.'[50] This was approximately the stage of development in labour policy reached in Canada during the 1941 Kirkland Lake strike. The conciliation boards could not enforce their decisions, and delay played an important role in undermining union organizing. Employers were not supposed to discharge employees because of their trade union activity, but there was no effective remedy if they did so.

In the mid 1930s, Senator Wagner of New York State proposed a labour statute that would have adequate means of enforcement. The result was the National Labor Relations Act of 5 July 1935, which established the National Labor Relations Board (NLRB) as an *independent* agency separate from the Department of Labor. The Wagner Act established a comprehensive framework for collective bargaining, and represented a legislative endorsement of its principles. Previously the government position had been equivocal. Disputes between employers and unions were regarded as essentially a private matter. While it was lawful for workers to organize and bargain collectively, it was equally lawful for employers to refuse to recognize unions and to discharge workers who joined them. The courts 'impartially' considered employee organizations and employer cartels as conspiracies in restraint of trade prohibited by the Sherman Act. *This legal system under which the government played a neutral role had the effect of tipping the balance of bargaining power in most American industries in favour of employers.*[51] The Wagner Act recognized this situation and remedied it. Similarly, in Canada during the early 1940s, Prime Minister King purported to maintain government 'neutrality,' while trade unions demanded a 'Wagner Act.'

The Wagner Act sought to achieve equality of bargaining power by balancing the organization of employers with the free exercise by employees of the right to bargain collectively through representatives of their own choosing. It was argued that inequality of bargaining power depressed wage rates and contributed to recession. Denial of the right to bargain caused strikes that obstructed inter-state commerce. Both problems could be resolved by encouraging collective bargaining, which would redistribute income and stabilize the business cycle. Compulsory recognition would remove a prime cause of industrial strife and provide a mechanism for resolving labour/management disputes. Whatever the merits of this economic rationale, it is significant that the Wagner Act contained a positive declaration in support of collective bargaining. A similar affirmation did not enter Ontario legislation until 1970 or Canadian federal legislation until 1972.

The act outlined a series of 'unfair labour practices.' An employer could not interfere with, restrain, or coerce employees in the exercise of their rights, nor could he interfere with the formation, selection, or administration of a union. This clause eliminated employer domination of company unions, so that henceforth bargaining had to take place between *independent* bargaining entities. The NLRB banned such overt practices as espionage, professional strike-breaking, private police, and industrial munitions so that most violence ended in the conduct of industrial relations. The Wagner Act affirmed the principle of a single bargaining agency, and gave the NLRB the power to determine the appropriate bargaining unit, and if necessary to conduct representation votes to determine which union should represent employees. The right to strike was guaranteed and the agency was encouraged to develop expeditious machinery for resolving disputes.[52]

The Wagner Act laid the legal foundation for the conduct of industrial relations in the United States. It raised the status and acceptability of the trade unions, and thereby stimulated trade union growth. Other factors – including the CIO campaigns in the mass-production industries and the AFL response to rival unionism – had been important prior to the ruling of the

Supreme Court upholding the Wagner Act in 1937; however, the act allowed labour to consolidate its gains and stimulated further organization. Most importantly, for the purposes of this study, the act made recognition strikes both illegal and unnecessary. The board interpreted 'the duty to bargain' to mean that collective bargaining between an employer and a union with majority support was *compulsory*, although a first contract was not. The parties did not have to reach agreement but they did have to 'bargain in good faith.'

In representation cases, the board devised methods to determine majority support for a bargaining agent: evidence at hearings, petitions, membership cards, and elections. By 1939 the NLRB relied heavily on the secret-ballot election. The majority of *those voting* determined whether or not a union would be certified and what bargaining agent would represent them. This procedure became a contentious issue in Kirkland Lake, and continued to be so for decades thereafter. In the 1960s, some Canadian jurisdictions still required a union to win an *absolute* majority of possible voters before it could be certified as a bargaining agent. By 1941, in the United States, representation cases replaced unfair labour practice cases as the most common kind of case heard by the NLRB, and the representation procedure became the main route to union organization and collective bargaining.[53] In Canada, in 1941, these problems remained unresolved.

Almost all of the questions that had been raised and resolved in the United States in the 1930s were raised in Kirkland Lake in 1941. As Logan has noted, the strike involved

many of the issues which caused the confusion of the period, namely the disagreement over the bargaining unit, the uncertain status of the company union, the right of an employer to deny bargaining rights to a majority union, the need and method of taking a [strike] vote ... Here also was a head-on clash between a representative company and organized labour which involved the participation of a number of unions and a wide collection of union funds. It was as well a test of government interference without a sufficient definition of its role – a test which it met with no great credit to itself and little assurance of stability in industrial relations for the future.[54]

It was logical for the Kirkland Lake miners to advocate a 'Canadian Wagner Act' to the Canadian government.

THE WAR ECONOMY AND THE 'LABOUR PROBLEM'

The conversion of the economy to a war footing required unprecedented government intervention and regulation of economic life. Under the War

Measures Act (1939) the Canadian federal government assumed extraordinary powers to regulate labour supply and demand, the level of prices and wages, taxes, and the margin of profits. New government agencies were established to administer the new policies. The Wartime Prices and Trade Board (WPTB) was created in 1939 to control the supply of key commodities, but apart from imposing rent control in a small number of centres the board's early intervention in the consumer market was confined to a temporary freeze on the price of butter and bread.[55] After the introduction of complex wage-and-price-control policies, its powers and responsibilities were substantially increased.

Until October 1941 (a month before the beginning of the Kirkland Lake strike), federal government intervention in the economy involved tentative attempts to deal with specific pressure points.[56] Nevertheless, this limited intervention marked such a change from the pre-war situation that it was perceived as extensive. Thereafter (and especially when the second wage policy, PC 8253, was proclaimed in Octoberr 1941), government involvement in the economy became truly unprecedented.

When the war began there were still 900 000 registered unemployed in a work force of approximately 3.8 million. By 1941 Canada's labour surplus had been absorbed, and by 1942 there was a labour shortage. At the request of business and military leaders, the federal government introduced new manpower policies.[57] In March 1942 the National Selective Service (NSS) was established under the authority of the Ministry of Labour to control the labour supply, to distribute manpower more efficiently, and to increase productivity. Industries and firms were classified into four categories according to the kind of labour required and their importance to the war effort. In order to fill essential jobs, the NSS drew upon all possible manpower reserves and inaugurated a campaign to attract women workers who had not previously been employed.[58] The government actively intervened in the labour market to influence employment policies. No employer could advertise for employees except by arrangement with an NSS officer. No employer could interview or hire a person who did not possess an NSS permit to seek and accept employment. Seven days' 'notice of separation' was required before an employer could fire an employee or an employee could quit. After October 1942, 'labour exit permits' were required before workers could seek or accept employment in the United States. The explicit ban on 'poaching' of employees indicates the fierce competition for workers existing at the time.[59]

These conditions put upward pressures on wages as employers sought to attract or retain workers. Because of the tight labour market, workers were less afraid to join unions. If they were fired for union activity on one job (and despite the 1939 Criminal Code amendment making such management action illegal, it continued to go on),[60] they could find employment elsewhere. Full employment, abundant job vacancies, and an excess demand for labour relative to the available supply were very favourable conditions for trade union organizing.

At the same time there was more government 'control' of an employee's working life. The NSS[61] severely restricted a worker's freedom in the labour market, since a worker could be frozen in a job, transferred, or placed in a military training plan.[62] The essence of all of these policies was to treat labour as a factor in production that could be regulated by legislative and administrative fiat. As a result of this suspension of their individual autonomy and the increased regimentation of their working lives, workers turned to unions to represent them.

Until 1942 the manpower situation in the mining industry was as tight as in other areas of the economy. Miners had specialized training and were not easily replaced. Many miners from the Kirkland Lake area had left the mines to join the armed services,[63] and the mining companies were forced to hire more immigrants to fill the vacancies. The mine managers increasingly had to use the manpower available, even though they knew that there were union sympathizers among them. The presence of immigrants in the labour force provided the basis for the company charge that the organizing drive in Kirkland Lake and the subsequent strike were the work of 'enemy aliens.'

The labour market conditions produced higher wages, rising expectations, and demands for better working conditions. Wartime wages were high by Depression standards, although in 1941 most industrial workers were still not earning an adequate wage as defined by welfare agencies of the day. Some 32.1 per cent of wage earners made less than $450 a year, and only 6.8 per cent were earning over $1949 per year.[64] But the wage situation was improving. The recovery of the economy in 1939 resulted in a general increase in wages of one-half to one per cent. By 1940 this trend was much more pronounced, and the average wage rose 3 per cent. The increase in wages was highest in manufacturing because of the growing production of munitions.[65] This trend continued until 1943, although wage increases throughout the economy were not uniform[66] and were reduced by the government's wage control program. Those industries with the highest and most rapidly increasing wage rates were also the industries being unionized most quickly. Situations in which unions bargained wage increases successfully had a demonstration effect that encouraged other workers to form unions and demand more.

Wages were rising, but apparently not as quickly as the expectations of workers who were turning to unions to achieve their monetary goals. Their demands were a reaction to the tight labour market, the rising cost of living, and their experiences during the Depression, when for years they had had nothing. They were determined not to return to the circumstances of the 1930s. Their wage claims also reflected a general discontent with wartime working conditions. The standard work week varied from forty-eight to fifty-four hours,[67] but because of the war effort there was also a great deal of overtime. Assembly lines were speeded up to maximize production. Complaints were regarded as unpatriotic and most workers were just as interested in war production as was the general citizenry; however, they demanded that the regular work week be shortened to between forty and forty-eight hours, and that time in excess of that amount be paid for at a higher rate.

Pressure for higher wages increased with the rising cost of living. While inflation was more controlled during this crisis than during World War I, many older workers remembered the extent to which real wages had been undermined during that struggle and turned to unions for protection. After the proclamation of the wage control policy in 1940, workers' wages were subject to direct controls and most wartime wage increases were not increases in the basic wage rates, but rather legislated cost-of-living bonuses. The wage control order PC 7440 took the form of an instruction to boards of conciliation. The essence of the order was that basic wage rates in war-related industries were to be frozen at their 1926–29 level, or any higher level since established. Exceptions to the rule were allowed only when 'it can be clearly shown that when such level was established, wages were depressed and subnormal or unduly enhanced.'[68] As a partial concession to labour, conciliation boards were also allowed to authorize (not order) the payment of a cost-of-living bonus where purchasing power had been impaired. The bonus would be based on increases in the national cost-of-living index, and was to be paid in a flat amount per week to all the workers affected. These bonuses were intended only to help workers keep up with the rising level of wartime inflation. They were not meant to allow any increase in their basic wage rates. At first wage controls were to be applied *only* to the 'pressure points' in the economy where there was excess demand for labour. These were the areas where labour was most likely to win wage increases through collective bargaining. The wage policy was more stringently enforced as the war continued, and workers feared that even these wage gains would be rescinded at the end of the war and that thereafter there would be a return to unemployment and insecurity.[69] The establishment of a mechanism to protect them was paramount, and increased their support for union recognition. After the introduction of wage control, recognition became an issue in an increasing number of strikes.

Wage packets were further reduced by the 'voluntary' deductions for savings bonds, the Red Cross, and other war causes. Such deductions decreased workers' spending power and thereby fulfilled an aim of the government's anti-inflation policy. Workers also paid more direct and indirect taxes than before the war, so that while wages were higher, workers sometimes considered themselves materially worse off.[70] These insecurities prompted workers to join unions at a time when wages were strictly controlled, so that there was no guarantee of immediate economic benefit.

In Kirkland Lake, wages were better than in some other sectors of the economy, and probably could not have been raised substantially without a modification of the applicable wage controls. The mine owners pointed to rising wages and attributed employee unrest to ignorance, irresponsibility, outside agitators, and 'unpatriotic elements.' But the issue in Kirkland Lake involved recognition, not simply money, and it was for recognition that the employees struck.

The regulation of wages and working conditions by the political authority inevitably brought trade unions into politics and increased labour's criticism of the lack of labour representation on government policy-making bodies. Organized labour argued that the wage control policy was inequitable because it affected wage earners more than salary earners, and did not properly take account of low-wage industries and regions. There appeared to be a discrepancy between the sacrifices that labour and business were asked to make. Workers were continually told by management and government spokesmen that their work was important to the war effort, but rhetoric of this kind usually accompanied requests for further sacrifices. Even though there were price controls, it appeared that business was receiving generous subsidies for capital expenditures and was allowed to earn a comfortable profit. Early in the war and despite government pressure, business had refused to produce for government needs on the basis of a 5-per-cent profit margin.[71] Yet employee demands to protect their own position were labelled unpatriotic.

Labour was also concerned about the administrative organization of the war effort. The government met the need for experienced personnel by attracting large numbers of businessmen into government service. Their contribution to war production was undoubtedly significant, but they brought with them their antagonistic attitudes towards unions and collective bargaining.[72] Labour resented the anti-labour attitudes of the 'new men' drawn from business into C.D. Howe's Department of Munitions and Supply. There developed a deep distrust of the people entrusted with administering government policy – as is clear from the TLC and CCL representations to the government and the resolutions at the 1940 TLC convention.[73] The line between business and government became blurred as former employers became government officials and vice versa. While organized labour remained outside the formal power structure, business influence in government and society was much greater, and that influence was reflected in the government's labour policy.

In these wartime circumstances, trade unionism was more than a device for raising wages; it became a vehicle which focused broader social and economic discontent. The common effort industrial workers were making to wage war, their resentment over 'unequal sacrifice,' their dissatisfaction with wages, their insecurity about the future, their opposition to the imposition of wage controls containing inherent inequalities, their struggle against employers' refusal to recognize unions despite majority support, their lack of influence on the government and lack of status within the community all played a part in the Kirkland Lake and other wartime strikes. These factors led to trade union organization, industrial unrest, and ultimately to political opposition through support of the CCF, and help explain why the wage issue gradually gave way in importance to the issue of union recognition. The union was seen as a vehicle for redressing all of these grievances, and consequently there developed significant support for a non-monetary issue.

The new constraints, when combined with production pressures, long working hours, and labour's increasing resentment of the government's wage and collective bargaining policies caused the 'labour problem' to escalate during 1941 and 1942, and to reach explosive proportions by 1943. All Canadians were asked to make a contribution and sacrifice as equals in a war effort for democracy; but this only heightened the dissatisfaction with 'industrial autocracy.' As 'equal' participants in the war effort, industrial workers wanted equal rights on the job, in the economy, and in the councils of the nation. Strong unions were their vehicle to acquire those rights.

MANAGING THE 'LABOUR PROBLEM'

The government was so preoccupied with the 'labour problem' as it conceived it that little attention was paid to employees' desire for recognition or collective bargaining. Organized labour attributed the rising level of industrial unrest to the government's management of the war economy, and in particular, to its legislative and administrative responses in the labour relations area.

In 1939 when the War Measures Act made the federal government pre-eminent in labour matters, the government had no positive collective bargaining policy. Instead, it based its response to labour/management problems on the policy of the Borden government during World War I. At a time when thousands of employees were joining unions there was no legislative support for their endeavour, or protection should their employer make reprisals. The 1939 Criminal Code amendment was inadequate in this regard.

In 1939, the Industrial Disputes Investigation Act (IDI Act) was extended by order-in-council PC 3495 to all war-related industries – about 85 per cent of all industrial activity; however, it did not contain any provisions for union recognition and was primarily concerned with avoiding strikes through the process of compulsory conciliation, which became a necessary pre-condition for a lawful strike. Conciliation implied a 'built-in' compulsory delay that was particularly troublesome in recognition disputes where the timing of strike activity was crucial. Delay could interrupt the union's organizational momentum and give the employer the opportunity to relocate production, recruit strike-breakers, and promote management-controlled 'employees' committees' to compete for the loyalty of the work force and hinder the development of independent unions. The application of the IDI Act, therefore, handicapped trade union organization, and inevitably benefitted management.

The IDI Act took no account of the different types of industrial dispute. Disputes concerning union recognition and collective bargaining required different treatment from those involving wages and working conditions. The act proved 'unsuited to deal with disputes arising out of the refusal of the employer to recognize and deal with trade unions.'[74] The recognition of a union as the 'bargaining agent' for employees was essential to the establishment of a collective-bargaining relationship, but there was no certification mechanism in the IDI Act. The number of recognition disputes increased throughout the war, but because they involved the very existence of the union and the legitimacy of its activities, they were not amenable to mediation and compromise. The very existence of one of the parties was not an issue for which there was a 'middle ground.' The act's unsuitability for recognition disputes was not immediately foreseen because of its narrow application prior to the war. The act had not applied to industries where the mass of employees were being organized. Recognition strikes before conciliation were illegal, but conciliation provided no solution. Yet strikes were not acceptable to the public, which felt that industrial conflict was detrimental to the war effort. Management's approach to the problem was to undermine trade union activity. Businessmen encouraged company unions in the hope that they could curb unrest, give workers a form of 'consultation' with management, and prevent the development of independent trade unions.[75]

In order to deal directly with the increasing number of disputes in which employers refused to bargain collectively, the government could have enacted legislation similar to that operating in the United States. The labour movement needed something along the lines of the American Wagner Act, but at the beginning of the war the federal government had no intention of enacting a Canadian version. Instead, in June 1940, the government was persuaded (by a labour delegation)[76] to proclaim order-in-council PC 2685 (see appendix): a *declaration* of principles that the government hoped labour and management would adopt, and that would prevent industrial unrest, which might prejudice the war effort.

The order was an effort to furnish a *voluntary* formula for the resolution of recognition disputes. It urged employers to voluntarily recognize unions, negotiate in good faith, and resolve disputes by means of the conciliation machinery.[77] The government itself sought to maintain a position on collective bargaining that it characterized as 'neutral.' By its statutory silence it implied that labour/management relations were essentially a private matter. This ignored the fact that a legal system that restricted the right to strike, but in which the government played a 'neutral' role, had the effect of tipping the balance of bargaining power in favour of employers.[78] As J.L. Cohen, a prominent labour lawyer of the day, wrote: 'It [PC 2685] ignores the essential fact that in the main, employees are not free either to organize or to negotiate and that no legislative protection, whether the right to organize or to negotiate is furnished by the order in council, by Section 502A of the Criminal Code or by any of the provisions of the IDI Act.'[79]

Government policy was not neutral but rather sought to promote industrial peace at the expense of collective bargaining. Prime Minister Mackenzie King, for both philosophical and political reasons, disliked militant unions that might create conflict and preferred conservative unions or employees committees because they played an accommodative role. The idea of employee committees, as opposed to independent unions, had always appealed to King, who had designed and implemented such a plan in Colorado earlier in his career.[80] His ideas about trade unions did not change substantially throughout the years. In an early pamphlet, *The Four Parties to Industry*, he espoused the belief that the state was composed of the harmonious interests of labour, capital, management, and government. He wanted to see 'the establishment of permanent standing joint industrial councils, embracing all the workers and all the employers in a given trade or industry and concerned with the determination of industrial policies and the fixation of industrial standards ... enforceable by government.'[81] In the 1903 Royal Commission Report on rail and coal miners' strikes in British Columbia, King wrote that workers should show employers 'by experience that it is good to deal with unions as such and that unions will regard the *interests of their employer as their own.*'[82] He also believed that strikes 'solely because of [the use of] non-union labour ought to be made punishable by law' and that sympathy strikes should be repressed.

In *Industry and Humanity*, published in 1919, King further elaborated his concept of 'industrial democracy.' It was essentially a conservative one, which did not admit to the existence of employee interests that were different from, and might occasionally legitimately conflict with, those of employers. His was a model of natural industrial harmony because of a basic community of interest. He distinguished between 'good responsible' unions, and 'bad irresponsible' ones; he was not here referring to the desirability of the union being responsible to its members. Responsible unions were those that accommodated management; unions were irresponsible when they disturbed the underlying consensus that King believed 'should' exist on the industrial scene, and exacerbated conflict. During World War II, King maintained the same views about unions and their leadership. The CCL unions were seen by him to be irresponsible. The notion that employees, at least in the short run, had different interests from shareholders, and that accordingly a conflict of interest was inevitable, was entirely rejected.

King's views were not merely expressed in his writings, he also put them into practice. His Colorado Plan had been designed for the Rockefellers in order to alleviate industrial unrest by giving workers a sense of increased participation – but not union recognition. In a letter written to Rockefeller during World War I, King suggested that it was not participation in decision making that was important, but rather the economic standards that that participation produced. Enlightened employers who maintained adequate standards need not worry about union recognition:

Speaking generally, there is going to be a large amount of unemployment as a consequence of this war ... union recognition simply for the sake of recognition is going to be seen to be less pressing, as an immediate end, than that of maintaining standards already existing, and unions may *rightly* come to regard as their friends and allies, companies and corporations large enough and fair enough to desire to maintain these standards of their own accord.[83]

In King's analysis of labour relations, he was more concerned with the *results* of collective bargaining than the *process*. The focus was on the terms and conditions of employment established, not whether they were established unilaterally or through the process of bargaining. This may explain why as a prime minister with direct experience in labour matters, he was not concerned with establishing collective bargaining policy. Yet in Kirkland Lake, the desire for collective bargaining came to be *the* principal issue in the dispute.

The Colorado Plan itself did not 'democratize' the working environment, and the sharing of authority envisaged by the plan was carefully circumscribed in accordance with management's perception of its rights and interests. However, the plan did change Rockefeller's image from that of a 'robber baron' (after the Ludlow Massacre), to that of an 'industrial statesman.' (The Ludlow Massacre in 1914 was the bloody climax of a UMWA strike in southern Colorado against a Rockefeller-owned company; sixteen people were killed.) Through the plan, Rockefeller and King hailed a new partnership of capital and labour. In reality, the distribution of power in that situation, as in the Kirkland Lake situation in 1941, had not been substantially changed.

King's consistent aim in labour relations was always to avoid industrial strife and unrest. In the 1920s he felt it would further the ultimate development of responsible trade unions if employers became adjusted to employee representation in industry through joint industrial councils. Then gradually, as accommodation to this new relationship developed confidence and displaced suspicion and hatred, trade union organizations could replace the unorganized employee groups as the agency through which the workers obtained a voice in the determination of employment conditions.[84] Believing as he did that the interests of employees and employers were ultimately the same, he saw his role as providing mechanisms (such as conciliation) that would break down the barriers, purely artificial, between labour and management.[85] In that decade the Department of Labour endorsed industrial councils, and adopted a policy to encourage the development of such organizations, although there was no indication that the department ultimately sought to convert such councils into unions. The council development was to be *independent* of trade unions. Unlike the British Whitley councils, which declared that trade unions were the necessary basis upon which joint councils were to be established, the Canadian federal government encouraged councils as an *alternative* to unions.[86] The 'employees committees' concept introduced during World War II was a new version of the old council idea, based on the view that this kind of relationship between management and its employees would eliminate industrial conflicts.

The adversarial nature of industrial relations became apparent when unions organized aggressively during World War II, and management resisted this surge of organization. In the early years of the war, unions were changing the industrial status quo, but the federal Department of Labour refused to accept this situation until the twin realities of industrial unrest and political considerations forced a change in 1943 and 1944. Prior to that time the department continued to operate on the old premises. The presence of such ideas in the labour department explains why the management of the mines could agree with King's and Labour Minister Humphrey Mitchell's 'accommodative co-operative concept' of labour relations so that both the government and the employer could on the one hand accept employees' committees, but on the other hand reject union recognition. Neither group was prepared to accept that there were different interests in industry requiring representation, and that the recognition of the legitimacy of these interests was *the* issue. The government's view was that the assertion in industry of a separate, independent, and conflicting employee interest was illegitimate and designed its policy accordingly. The result was that in 1940, PC 2685 and not collective-bargaining legislation became the labour policy of the government.

The unenforced 'declaration of principles' became the focus of much bitter debate and contributed to labour's disaffection. Although there was little government concern for collective bargaining, wage increases in the war industries prompted the government to control prices and wages. In contrast to the wage control policy (PC 7440, December 1940), which was widely publicized and firmly enforced, the labour relations principles were ignored by employers,[87] and never followed by the government itself in industries under its own control. While the government was prepared to impose compulsory wage controls, compulsory conciliation, compulsory reallocation of labour, and later compulsory strike votes, it continued to maintain that its opposition to 'compulsion' precluded the introduction of collective-bargaining legislation.

The government refused to acknowledge the contradiction between its unenforced 'declaration' on collective bargaining on the one hand, and its *mandatory* policy on wages on the other. Even the wage order was ambiguous in its reference to collective bargaining. It suggested that collective agreements conform to PC 2685, but this in itself was meaningless. There would not be a contract if there were no recognition of collective bargaining.[88]

The labour movement initially believed that the wage order required adherence to PC 2685 principles. This was not what the government had intended, but it was slow to clarify this confusing situation and was unwilling to assume any clear, unequivocal position on collective bargaining. Ultimately, the government's interpretation of its wage policy negated collective bargaining. On 6 June 1941, the minister of labour finally admitted that PC

2685 was not mandatory and expressed only 'an opinion of the government.'[89] This statement ended any possibility of adherence to the declaration of principles for it meant that the government would refuse to impose them on unwilling employers.

It was evident from the structure of the new administrative agencies that labour relations policy had a low priority. Between June 1940, when the government proclaimed its declaration of principles, and December 1940, when the first wage policy was proclaimed, the government was preoccupied with securing the agreement of the provinces on the unemployment insurance program that had been recommended by the Rowell Sirois Commission Report (August 1940). There was greater concern with the problem of labour supply than with labour relations policy: 'The Prime Minister viewed an unemployment insurance program as politically imperative and a policy of labour supply as equally imperative in order to meet production and military targets. No similar pressure existed for legislation on collective bargaining, or on the status of trade unions.'[90] The Economic Advisory Committee – the principal committee advising the government in 1940–41 – drew its members primarily from the finance department. The principal decision-making mechanism of the Cabinet – the War Committee – excluded the labour minister. In June 1940, the government established an *advisory* body, the National Labour Supply Council (NLSC), to advise the minister of labour on any matters related to the supply of labour. Labour had been promised representation on this body, and the NLSC was, in fact, composed of representatives of management and labour organizations. It existed until February 1942 when it was replaced by the National War Labour Board (NWLB). Labour remained dissatisfied because it believed that advisory committees like the NLSC were ineffective and that the real decisions were made elsewhere.

In October 1940, the government established the Committee for Labour Coordination, which was composed of senior civil servants. With the exception of the prime minister and the War Cabinet, it exercised the most direct influence on the development of labour policy. There was no direct labour input.

THE 1941 CRISIS

The IDI Act and wage control became the chief instruments of industrial relations policy; increasingly the labour movement became dissatisfied with that policy. The IDI Act restricted the right to strike and PC 7440 and successive wage policies restricted the wage levels for which employees could bargain. Neither contained any positive measures to promote collective bargaining or gave substance to the government's declaration of principles. Early in the war, the labour movement tentatively supported wage controls. In the face of adamant government support for the idea as a necessary complement to price control, labour did not want to be blamed for sabotaging the war effort.[91] The government had presented the wage order as a *fait accompli*. After watching the application of the policy in specific situations – particularly at the Peck Rolling Mills plant in Montreal – this support changed to opposition. Existing inequities were frozen into the wage structure and inconsistencies in the application of the wage policy emerged. In a number of cases conciliation boards withdrew from disputes when they were settled, even if the settlement violated the wage control principles. Some war industry agreements contained significant wage increases. The government's wage control policy was in a shambles: 'Its only real effect was to provide employers who wished to resist wage demands with an elaborate rationale.'[92]

Towards the end of March 1941, the minister of labour's interpretation of the wage control order became a further source of labour dissatisfaction. In January, the minister asked the NLSC to comment on the recommendations of the Committee for Labour Coordination for additional wage restrictions. The NLSC rejected this recommendation and suggested that when the order was raised before a conciliation board, it 'should be left to be interpreted by the board on broad lines of justice and that wage rates and cost of living bonuses should be adjusted ... having in mind the particular conditions applicable in each case.'[93] Despite this advice and Cabinet meetings with both labour federations, the minister issued a series of interpretations called the 'Wartime Wage Policy.'

The new interpretation increased the rigidity of the controls. On 24 April 1941, CCL President Aaron Mosher resigned from the NLSC in protest. He was opposed to an inflexible policy formulated in Ottawa and applicable to every situation throughout the country and believed that conciliation boards should be free to make recommendations about wage rates and bonuses in the light of circumstances in a particular case. Since the principle of control had originally been endorsed by the NLSC, Mosher did not want his membership on that body to imply that he concurred in the department's subsequent interpretation.

The real problem was that a policy based on 1926–29 wage levels did not reflect an expanding labour movement and collective-bargaining system. In specific collective-bargaining situations, the interpretation of the order had frustrated employee demands. The wage policy had resulted in the application of a fixed formula that ignored the merits of the cases of different groups. This was the antithesis of collective bargaining and could not accommodate the particular needs and circumstances of the bargaining parties. Employee dissatisfaction was inevitable under these conditions, since wages were determined in accordance with decisions made in Ottawa over which they had no influence. At the same time, employees remained subject to the authority of management. There was no participation or sharing of power in either government or industry. The labour movement believed that the collective-bargaining and wage policies were related. Unions were prepared to sacrifice increases in wages so long as the government guaranteed collective-bargaining rights to give workers some form of protection from arbitrary employer action.[94] Mosher's resignation led to a reconsideration of labour policy and some cosmetic changes. The Industrial Disputes Inquiry Commission (IDIC) was created, and an amendment to the IDI Act ensured greater impartiality of conciliation boards.

The third change was more significant. PC 7307[95] added another prerequisite to legal strike action. Following the report of a board of conciliation, no strike could occur until a strike vote had been taken. A majority of all employees, who *in the opinion of the minister of labour* were 'affected by the dispute or whose employment might be affected by the proposed strike,' had to endorse strike action. The Kirkland Lake strike was to be the *first* dispute to which this order applied.

In August 1941, the Cabinet endorsed in principle a new price-and-wage-control policy to ensure greater wage stability and more effective administrative machinery. In October PC 8235 – the Wartime Wages and Cost-of-Living Bonus Order – was proclaimed. The National War Labour Board (NWLB) replaced conciliation boards as the mechanism for applying the wage order. Five regional war labour boards were created with a right of appeal to the NWLB. These were later increased to nine, one for each province.

The wage control program was extended to cover all the employers operating in the federal wartime jurisdiction. Employers were prohibited from raising wages without the permission of the new boards, so that in effect there was a wage freeze. In addition, the NWLB was empowered to assist in the development of new wage and labour policies. It could make recommendations to the minister and advise him on many other labour-related questions. In effect the NWLB would supplant the NLSC: 'The Prime Minister and the Minister of Labour both felt that this concession had to be made to labour, in lieu of the lack of government action on a Canadian Wagner Act, and to satisfy demands for representation.'[96] There was disagreement over another 'concession' to organized labour – the extension of the cost-of-living bonus. Both the Committee on Labour Coordination and Dr Bryce Stewart of the labour department advised the Cabinet that organized labour would reject the new order, unless the cost-of-living bonus was extended along with wage control. Although the minister of finance, Mr Ilsley, believed that this aspect of the policy was inflationary, the Cabinet, urged by the prime minister and labour minister Norman McLarty, eventually supported the extended bonus.

Between August and October 1941, labour leaders were not consulted about the new wage-and-price-control policy. The prime minister was conscious of labour's sense of exclusion and on 15 October he met with various labour leaders. He submitted the wage order to the NLSC the next day and addressed the nation on radio before the new order was announced.[97]

At the meeting with the labour leaders, King asked for their support 'in these perilous times,' and assured them that he would shortly put a 'labour man' in the post of labour minister. He requested that the labour leaders refrain from opposing the order publicly even if they could not support the measure. He acknowledged that the development of wartime labour policy had not been altogether satisfactory, but promised that in future it would be revised in co-operation with representatives of employers and employees and the provincial governments.

King recorded his impressions of the meeting in his diary. He noted[98] that the labour leaders

had expressed an entire lack of confidence in the Government in the matter of its labour policies. They could not understand why we had not enforced our own policies ... with respect to collective bargaining ... I had to confess that, as Prime Minister ... I shared their impatience about the lack of readiness in carrying out the Government's policies, but hoped that the step we were proposing to take in constituting a National War Labour Board and 5 regional War Labour Boards might help to overcome that failure.

King believed that, with the exception of the CCL representative, he had won the confidence of the labour leaders: 'I felt ... we had antagonized labour instead of holding them with the Government as they should be ... The day was certainly saved by meeting the labour men ...'

King was being overly optimistic. The labour leaders were not convinced that he was sympathetic to their position. He had conceded very little except that the NWLB would advise the government about a *future* labour policy. At the same time he had devised and imposed an extension to the wage control policy, without consulting labour's representatives. As a result, the labour movement had little faith in King's promises.[99]

By the fall of 1941, both the TLC and CCL conventions openly criticized the government's labour policy. Dissatisfaction focused on three issues: the inequity of wage controls (particularly as demonstrated in the dispute at Peck Rolling Mills in April 1941); the failure of the government to involve labour on decision-making bodies; and the absence of a Canadian Wagner Act, which would guarantee collective bargaining. The TLC endorsed the principle of wage controls, but not the restrictive interpretation of the wage order. The CCL called for the repeal of wage controls. Labour's dissatisfaction over its lack of representation in government increased as it became evident that the NLSC was becoming a 'rubber stamp' for policies that were developed and approved by the government before being referred to the NLSC. The TLC objected to the failure to apply the principles of PC 2685 in government-controlled industries. The CCL's demand for a Wagner Act received more widespread support from its membership than did the similar TLC resolution. This is not surprising, since its member unions were more directly affected by the lack of legislative protection for collective bargaining. The most notable case involved a dispute at the National Steel Car Company in June 1941.

Labour's criticisms of government policy were effectively summarized by Cohen in his book *Collective Bargaining in Canada*, published in 1941. This book attributed labour unrest to the resistance of employers to employee organization. Cohen severely criticized the government for its 'neutral' position, which, in his view, aided anti-union employers. He argued that collective bargaining was in the public interest: 'freedom of association and the establishment of collective bargaining are not the expression only of civil rights of workers but of social and industrial functions which are basic and essential in a well ordered society.' Government was avoiding its responsibility. It preferred to 'appear to be filling the role of umpire between competing social forces and behind that role ... to conceal the fact that as a government it has failed to discharge its primary duty of prescribing the rules. Umpiring without rules is a makeshift process and that in great measure marks the whole attitude of government today on the question of labour relations and collective bargaining' (p. 15).

Cohen's summary of labour's critique of the government's policy was widely publicized. Privately the prime minister admitted that his criticism was largely valid. King was well aware of labour's major criticisms and demands.[100] At the same time he was also receiving representations from various individuals, business groups, and other associations, advising him to abolish the right to strike, to intern labour leaders, and to limit the organized labour movement. In June 1941 the Canadian Manufacturers' Association

(CMA) held a three-day conference on the labour situation,[101] and its president subsequently told the prime minister that the labour situation was rapidly deteriorating because of 'sinister groups of dangerous men' who were trying to 'seize control' of the key war industries. The CMA's view was shared by many businessmen, including the gold mine operators.[102] Ontario's Premier Hepburn joined the business interests in charging that labour was falling under the control of 'sinister foreign influences.' He campaigned publicly against CCL unions, both prior to and during the Kirkland Lake strike. The state of labour/government relations steadily deteriorated.

Unlike the business community, which was virtually running the production effort (and reaping considerable benefits), the labour movement remained unrepresented and largely unheard, except when through the exercise of industrial strength a government economic objective was jeopardized. Because labour was excluded from the formal decision-making process, its opposition to the government was expressed only on the industrial scene. In order to understand the intensity of this labour opposition on the one hand, and the insensitive attitude of government administrators of labour policy on the other, it is necessary to examine the government role in several key strikes that preceded the one in Kirkland Lake. Each involved an important element of the government's labour policy, and each contributed to labour's alienation. Eventually this alienation, particularly over the collective-bargaining issue, peaked in the Kirkland Lake strike and prompted the two major labour federations to adopt common legislative goals, to forge new political alliances, and to engage in overt political opposition.

The dispute at the National Steel Car (NASCO) plant in Hamilton in the summer of 1941 was a recognition strike occurring immediately before the Kirkland Lake strike. When the company refused to meet, the Steelworkers' Organizing Committee (SWOC) applied for a conciliation board. The conciliation board recommended that a plant-wide, government-supervised representation vote be conducted, and if the union won, the employer should begin negotiations. The union accepted the report but heard nothing further for a month. On 29 April 1941, the membership unanimously voted to strike. Immediately the government became concerned about the possible disruption of war production and responded within *two days* by appointing Ernest Brunning, controller of the plant, to implement the conciliation report. The representation vote was taken and the union won, but the controller refused to meet. Significantly, he advised the conciliation board that he was pursuing this course of action in accordance with instructions that he had received from the government.[103] The absurdity of this situation was noted by Cohen, the labour nominee of the conciliation board, who wrote to the government:

'There appears to me to be something incongruous in the suggestion that a government-appointed Board should be required to inform a government-appointed Controller that the principles and policy of an order-in-council [PC 2685] enacted at the behest of the government appointing both the Board and the Controller should be observed and lived up to.'[104]

Brunning called upon the employees to appoint 'a representative committee' to meet with him and consider *his* proposals regarding hours and wages. These proposals were implemented a week later. This procedure was contrary to the principles of collective bargaining embodied in PC 2685, but it was obvious that the government was not going to enforce its order. In July the union called a second strike. After mediation, the strikers returned to work on the understanding that negotiations would finally begin. No negotiations took place, but the controller announced that the workers would be 'free' to join any union or employees' association of their choice. He obviously preferred to deal with the association that he himself had established and encouraged in the summer of 1941. Eventually a new controller negotiated collective agreements with both the union and the employee association. Despite the representation vote, the union had not achieved official recognition or the status of exclusive bargaining agent. The two organizations vied with each other until the United Steelworkers of America was finally certified in September 1945. The conduct of the government and its appointee created considerable disillusionment within the trade union movement. Not only was the government unprepared to support union recognition and the principles of PC 2685, it also had condoned the establishment of an employer-dominated committee that had been used to undermine the existing union.

Concurrent with the NASCO dispute, the first major dispute concerning the application of the government's wage control policy arose at the Peck Rolling Mills plant in Montreal. Peck Rolling Mills was a wholly owned subsidiary of Dominion Steel and Coal Company (DOSCO). SWOC had organized 93 per cent of the work force and was granted recognition by the company on the recommendation of the conciliation board. The board also found that 50 per cent of the workers received less than 30.7 cents an hour. In addition to poor wages, the Peck employees worked long hours (50 to 80 hours) in substandard working conditions. The parties fundamentally disagreed on both the level of wages and the proper interpretation of PC 7440, the wage control order. The majority and minority reports of the conciliation board reflected this disagreement. The employer and majority report interpreted the order narrowly and concluded that within the terms of the order, the wages paid were 'fair and reasonable' and not depressed. The Peck wages were comparable to other industrial rates in the Montreal area. The majority report concluded that the most recent wage order, freezing wages, precluded it from recommending a raise, even though it recognized that the wages were inadequate.

The union and the minority report supported a broader interpretation of the order and argued that Peck employee wages should be compared to wages of other workers across the country engaged in similar work. In its view the Peck wage rate was 'depressed and subnormal' and could be adjusted under the provisions of the order. The minority report contended that the government's wage policy was aimed solely at preventing wages that were already reasonable from rising unduly; the order was not intended to freeze inadequate wages. There was nothing in the Montreal cost of living or in the company's ability to pay that justified abnormally low wages in comparison with those paid to other workers in the same industry. Accordingly, the minority recommended an increase in the basic rate to 40 cents an hour.[105] The positions of the parties and the proceedings of the board were closely monitored, for labour feared that the Peck case would become a precedent for other conciliation boards handling wage disputes, as indeed it did. In addition, SWOC was beginning to formulate its demands for a general basic wage increase throughout the steel industry across the country.

In April 1941 the Peck workers struck for 40 cents an hour. The government sought to persuade them to return to work without giving in to union demands.[106] Ultimately the Peck employees received an increase in their basic wage rate when the federal government, avoiding any direct reference to the dispute, increased the minimum wage for men to 35 cents an hour and for women to 25 cents an hour.[107] The employees returned to work and the dispute ended with the temporary collapse of the SWOC local. The inequities of the wage policy, the rigidity with which it was applied in the Peck dispute, and the inconsistencies with which it was applied elsewhere[108] increased labour alienation.

Finally, in July 1941, the Arvida incident occurred. Labour/management relations had been deteriorating at the Aluminum Company of Canada plant in Quebec for months. The plant was crucial to war production, so that when a sit-down strike occurred, the government believed a crisis had arisen. C.D. Howe, the minister of munitions and supply, was incensed and believed that the plant 'had been seized by a group of men led by an enemy alien ...'[109] Notwithstanding the jurisdictional problems, he wanted federal troops to be sent in immediately and threatened to resign if his demands were not met. As a most powerful member of the Cabinet, with primary responsibility for war production, his belief that labour organizers were aliens and saboteurs was an influential view within the government.

The prime minister suspected that the situation at Arvida was a simple labour/management dispute, but he tried to accommodate Howe, whom he viewed as indispensable to the government. The federal government could not send troops unless the provincial authorities requested them to do so. King phoned the president of the company and suggested that he request the provincial government to evict the men.[110] Thereafter troops were sent, but not before aluminum had hardened inside the pots in the plant and caused considerable damage.

In subsequent Cabinet discussions the Royal Canadian Mounted Police (RCMP) were instructed to investigate the men who were responsible for the sit-down strike and to arrest any subversive elements. The entire matter was to be investigated by a royal commission. Also, an order-in-council was passed giving the minister of munitions and supply the discretion to call out troops to assist the police in the event of whatever he considered to be sabotage in war production plants. The labour movement did not condone sabotage, but was alarmed that such unfettered power would be given to a man so obviously opposed to trade unions and collective bargaining; labour worried that Howe would make little distinction between their activities and those of actual spies and saboteurs.

When the royal commission on the incident reported, it found no evidence whatever of sabotage by enemy aliens at the Arvida plant. The strike was the result of a breakdown in relations between employees and management.[111] Labour's view of the situation had been officially vindicated. However, the order-in-council concerning the use of troops was *not* rescinded when it was established that the allegation of sabotage was unfounded.[112] This incident represented another breach of faith between labour and government, forging another link in the chain of labour discontent.

The experience of organized labour in these strikes, just prior to the Kirkland Lake strike, suggested that the government could not be trusted, that collective bargaining was threatened, and that union organization and growth would be actively discouraged by the government. The Kirkland Lake strike occurred at a pivotal time, when labour was ready to unite and fight back against these trends.

2

The local context: the community and the parties to the dispute

The Kirkland Lake strike became a national confrontation between the federal government and the labour movement over the issue of collective bargaining, but locally it was a protracted and extremely acrimonious battle between the mine operators and their recently unionized employees. The unusual bitterness of the local dispute and its aftermath can be attributed to: a) the background of the community; b) the nature of the gold-mining industry, and the consequent paternalistic view of labour relations that prevailed among the mine operators; c) the character of the work force; and d) the history of the union in the area.

THE KIRKLAND LAKE COMMUNITY

The battleground of the 'big strike,' as it was later remembered, was the small northern town of Kirkland Lake, which had developed as a one-industry town, dependent almost entirely on the mining industry. Historically such one-industry towns have been particularly prone to violent strikes[1]; but they have also been the scene of excellent industrial relations.[2] Such communities seek to avoid industrial conflict because in a small, interdependent community where relationships are close and personal, the destructive effect of conflict is accentuated. When industrial conflict comes, however, the polarization and hostility surrounding such conflict is heightened by the fact that the community has a recognized, stratified, and inflexible social structure that is closely identified with the hierarchical structure of the company itself. There develop two completely divergent views of the process and significance of unionization, and as a result the strike inevitably divides the town into contending factions.

Communities like Kirkland Lake, which are isolated and dominated by a single industry, develop their own codes, myths, heroes, and social standards. One cannot visit this small town without hearing of Charlie Chow – a hotel owner who allowed prospectors to pay for meals with mining shares and who died a millionaire. Another folk heroine is Roza Brown, an old woman (also reputed to be rich) who built a shack on Government Road and tried to avoid taxes by presenting her abode to the British Crown. Such romantic characterizations become a part of the town's unique traditions and influence the attitudes of its citizens. S.A. Pain has described the Northern Ontario mining communities as towns born of 'high hopes and poverty; ... the search for an elusive fortune attracted rich and poor, expert and amateur, to scour the bush for individual Eldorados.'[3] The qualities most admired in these communities are rugged individualism, determination, eccentricity, success, wealth, and luck. The heroes are the underdog miners who overcome all obstacles to create a successful mining operation and make a fortune. These qualities were not only admired but also believed by the mine operators to be the key to their own success.

Kirkland Lake developed as a result of the discovery of minerals in Northern Ontario just after the turn of the century. Unlike Cobalt, which sprang up overnight, Kirkland Lake's development was hesitant and uncertain: 'Kirkland Lake ... struggled through anxious and difficult times to become one of the leading gold fields of the world.'[4] Its growth and decline parallelled that of the mining industry. The town originally emerged from a camp of bunkhouses erected for the expanding labour force, and throughout its history remained a creation designed primarily to serve the mining industry.

Serious exploration in the Kirkland Lake area (as it was named in 1907) began in 1906. The claims staked in 1911 by William Wright, later vice-president of the Wright-Hargreaves mine, and Harry Oakes, an 'experienced hard-bitten mining man' recently from Alaska, 'set the camp off with a momentum that ensured the development of other mines in spite of the dislocation of the war years that were soon to come.'[5] Oakes' claims were staked in what became 'the main break' of the camp at Kirkland Lake.

During World War I, the camp had only one mine in production, the Tough-Oakes Gold Mines Ltd, which had been incorporated in 1913 and later became the Toburn mine. During the same period, however, other mining properties (which later became the Teck-Hughes, Wright-Hargreaves, and Sylvanite Mines) were being constructed for future production. The first town facilities grew up around these mining areas: stores, a hotel, a stable, and the first post office opened in 1913. Early communication systems were developed with New Liskeard and Swastika, two nearby mining towns. Swastika was an important centre because the rail line went through it: 'With all trains stopping, Swastika was a busy place and mining men on the move could get out and collect the news from the usual train watchers on the platform or across the road at Boisvert's Hotel.'[6]

Haileybury became an affluent residential town where engineers, speculators, businessmen, and a few mine workers took up residence. In these war years prohibition was imposed in the camps, though whisky was smuggled in by rail to supply a lucrative bootlegging trade. Since Haileybury was not 'dry,' it became the liquor outlet for the camps. It had hotels on every corner and in the 'busy bar rooms of the hotels, spectacular mining deals were settled and fortunes won and lost.'[7]

Kirkland Lake continued to expand in the post-war period, as did the mines. In the 1920s the Lyric and Strand theatres were built, and in later years, the union was to hold membership meetings in them. Reeve Carter (the reeve between 1935 and 1942, when the strike occurred) had his first terms as reeve in 1926, 1927, and 1928. He had opened a drug-store in the town in 1923 to service the growing number of doctors who had been recruited by the mining companies.

With the growth of the population from 600 people in 1919 to 3000 in 1926, churches of many denominations sent preachers to the area and ultimately built more permanent structures. They formed an important part of the social life of the small community. The Roman Catholic Church was especially successful. 'The priests of the Roman Catholic Church had been active from the first, travelling up and down the new railroad and visiting logging camps, mine bunkhouses and boarding houses.'[8] In 1916 the (Roman Catholic) Church of St Jerome was built. Many miners were Catholic and other parishes grew. By 1932 the Church of the Holy Name was built – later to be attended largely by Ukrainian and Polish members of the community. The Church of the Assumption was attended primarily by French-Canadian Catholics. Protestant denominations were also represented, although for several years the first Presbyterian ministers had no church. The Anglican Church (St Peter's) was built in 1925; the Salvation Army arrived in 1925; and the Baptists built a church in 1927. In 1927 a synagogue was also erected on Station Road.[9]

In the 1920s municipal government offices expanded, as did the courts. Two local justices of the peace and a travelling magistrate were made responsible for the administration of justice. These sections of the community grew, but they continued to have financial or personnel connections with the companies. Ultimately all such developments were dependent on the continuing prosperity of the mines.

The 1930s was the boom period for the mining companies and for the town. With the influx of unemployed workers seeking employment, Kirkland Lake 'was one of the liveliest and busiest places in all Canada.' The mines were well established by this time and were prospering despite the general economic slowdown. In the context of the Depression, 'the most dependable things in the country were the payrolls, expenditures, taxes and individuals from the mines.'[10] This undoubtedly reinforced the mine owners' view of themselves as exceptional men who were being rewarded for their industry and business acumen.

The development of the town continued at an accelerated pace in order to meet the needs of its growing population. At first, makeshift housing accommodation and soup kitchens were the only facilities available to the men. Later, new hotels like the Park Lane and Princess were built and were filled at all times. New offices and stores went up. Kresge's built a store in 1938 and Eaton's established a branch in 1939. In 1939 William Wright donated $100 000 to extend the town's hospital facilities. Despite this extensive commercial development during these boom years, the influx of so many young men ensured that the town still retained some frontier characteristics. Ash's Hotel, known as 'Ashcan Hotel,' was at the end of the town and men could rent a room for an eight-hour period. The miners would also line up in front of the bars till 3:00 AM. The bars closed at 12:00 PM, and then the bootleggers took over.

World War II brought a new challenge to Kirkland Lake and to the gold-mining industry, which reached its peak of production and profits in the early part of the war. In 1942, when it was declared a 'non-essential' war industry, decline set in. This change had a profound effect on the course of the Kirkland Lake strike, its aftermath, and on the town itself.

The mines had always been the largest taxpayers in Kirkland Lake. As the output and profits of the mines dropped in the period of 1941–42, so did the level of municipal taxes. As the mines reduced production, they also reduced the size of their work force, resulting in a large-scale exodus of population from the community. The insecure future of the mining industry meant an insecure future for the town: 'As the mines progressed from a slow start to a triumphant climax, so the town rose from a muddy street of shacks to a smart centre serving a wide area.'[12] Similarly, as the gold-mining industry was given a lower priority in the economy, this change immediately affected the community.

THE NATURE OF THE GOLD-MINING INDUSTRY
AND MANAGEMENT ATTITUDES

Both the state of the gold-mining industry in 1941 and the managerial attitudes developed over the years contributed to the companies' hostility to
employee organization and to their adamant refusal to recognize the union.
Because of the way the industry had developed, the mine operators were
accustomed both to success and to being in control. Growth had been steady.
The major gold strikes were made in 1907. By World War I, one mine was
operating and several other mining properties were being prepared for future
production. Mining development continued throughout the war. Though the
first rich discoveries had been exhausted by 1918, profitable ore was known
to be there. By that time Kirkland Lake was 'the richest camp in the world.
One or two camps were handling richer ore but not in such great quantity
and other camps handled far greater tonnages of ore but less than half the
grade.'[13]

The 1920s was a period of financial consolidation for the mines and for
overcoming difficulties in mining the ore. Unlike many other gold fields, the
gold of Kirkland Lake was very deep underground. Consequently, the mines
had to expend capital and develop special technology in order to mine such
gold profitably. These technical difficulties were overcome and the expansion in the number and scope of mining operations in this period meant that,
by 1930, the Kirkland Lake camp had come to occupy a position of real
importance in the mining world.

During the 1930s Depression, the gold-mining industry boomed, and gold
prices rose from $20.67 an ounce to $35.00 an ounce on the international
money market. In 1940 Canadian gold was selling at $38.50 an ounce
because of the devaluation of the Canadian dollar. In the Depression decade,
production rose rapidly to a new peak – two million tons a year in 1936,
1937, and 1938. In 1937 the mines paid $20 260 000 in dividends.[14]

In the 1930s, the gold-mining industry made important media and political
contacts and became a powerful business lobby. These contacts played an
important role during the 1941 strike. It was in this period that William
Wright, founder of the Wright-Hargreaves mine, met George McCullagh. 'It
was a case of brains meets money and Bill's money was put to work.'[15] This
business combination refinanced the Bidgood mine. More importantly, in
1936, Wright (at McCullagh's urging) bought two Toronto newspapers, the
Globe and the Mail and Empire. McCullagh merged these newspapers to
form the Globe and Mail. As McCullagh said of the purchase: 'I thought I
could do something for the country by making our mining industries better

known. Anything that is of advantage to mining is of advantage to the country as a whole.'[16] Not surprisingly, the *Globe and Mail* was the most 'pro-management' daily newspaper in the country during the Kirkland Lake strike. The role of this paper, as McCullagh and Wright had planned it, was to act as a spokesman and protector for Canada's burgeoning mining industry. This was the function the newspaper performed during the union's organizing campaign, and during the strike.

In the 1930s McCullagh developed important political connections, and he became close to Mitchell Hepburn, the premier of Ontario. He supported Hepburn in the 1937 election, which was fought more against the CIO than against Earl Rowe's Conservatives. McCullagh's personal friendship and social and financial relationship with Hepburn undoubtedly influenced the premier's ideas about the CIO and his decision to intervene in the 1937 Oshawa strike. The intensity of anti-union militancy on the part of Canadian employers increased after 1937.[17] In Kirkland Lake, Mine Mill Local 240 faced a particularly intransigent group of employers. Their friend, Premier Hepburn, was adamantly opposed to the CIO. In 1937 Hepburn had announced that he would call in the armed forces 'to protect the mines of Northern Ontario from the CIO.'[18] He had proclaimed: 'Let me tell [John L.] Lewis here and now that he and his gang will never get their greedy paws on the mines of Northern Ontario as long as I am Prime Minister.'

When two of his cabinet ministers (David Croll and Arthur Roebuck) resigned, he wrote: 'if the CIO wins in Oshawa it has other plants it will step into ... it will be the mines, [it will] demoralize the industry, and send stocks tumbling.'[19] As a stockholder himself,[20] Hepburn was not anxious to see this result, nor did he wish to disappoint his friends and supporters in the mining community who believed that a union victory in Kirkland Lake would encourage union organization elsewhere in the province. By 1941 McCullagh had become a pro-conscription Meighen Conservative (as were most of the operators), and had lost faith in Hepburn. However, his earlier influence on Hepburn's ideas and the premier's close relations with the mining industry prevailed in the Kirkland Lake strike. The mining industry continued to have an important ally, and that ally provided massive police intervention when requested to do so.

World War II brought a new challenge to the Kirkland Lake gold-mining industry. During the union's organizing drive and the resulting strike, the high production and the increased price of gold contributed greatly to Canada's purchase of war materials. Gold production in the first two years of the war was 'an essential element in our efforts to strengthen the Canadian exchange position.'[21] The government recognized this situation by designating gold mining as an 'essential' war industry. The industry was given priority in the maintenance of manpower and in the purchase of materials. Nevertheless, there were complaints by mine operators about the number of skilled men leaving the mines to enlist in the armed services.[22] This special status for gold mining had an impact on the industry's industrial relations, for management could plausibly argue that it was 'unpatriotic' to strike. The union argued that if gold miners' work was essential, then remuneration should be improved and that better production would result.

The opening of hostilities in 1939 occurred at a high point in the development of the gold-mining industry, which had been expanding steadily since the rise in the price of gold bullion to $35.00 an ounce in 1934.[23] In 1939 the mines were enjoying the benefits of this expansion; gold mining was a leading industry when the war broke out. In 1939 the output and profits from the gold mines reached new levels, only to be surpassed in 1940.

The peak of gold production and gold profits occurred in 1940. In that year Wright-Hargreaves, Sylvanite, Toburn, Kirkland Lake Gold, and Macassa mines attained their maximum levels of production. Lake Shore and Teck-Hughes were past their peak, although they were still very prosperous. The combined production of the mines in 1940 was $32 million.[24] As yet, the demands of the war economy only slightly affected the industry, and, because of the war, the mine operators had decided to mine more low-grade ore in order to achieve higher production. This was an expensive process, but still a profitable one. The mines also continued exploration for new ore bodies.

By 1941 the adverse impact of the war began to be felt. Despite the increase in production, profits began to decline. The mining of low-grade ore raised operating costs. The cost of supplies was higher because of scarcity and wartime inflation. Many of the best workers had left town. Management, in accordance with government policy, had had to pay cost-of-living bonuses to its employees to help them maintain their purchasing power in a period of rising prices. The level of productivity had decreased slightly, a phenomenon management attributed 'almost entirely to CIO agitation, enlistment of better class men and labour shortage due to the demands of the war plants.'[25] Most importantly, the government raised taxes as part of the 'industry's contribution' to financing the war. The government wanted to avoid a repetition of war profiteering that had been politically unpopular during World War I, and in the latter part of 1940 it levied a new excess profits tax. After 1940 all mines paid in taxes a minimum of 30 per cent of their total profits. This included an 18-per-cent corporate tax and a 12-per-cent excess profits tax.[26]

This policy represented a great change in an industry that previously had been taxed very little.[27] The response of the mining industry was immediate:

several mines cut their dividends to shareholders and eventually almost all of the mines reduced their previous efforts to increase tonnage. Further development schemes were postponed.[28] Beginning in 1941, the mining industry complained bitterly about high taxes in the columns of the *Northern Miner*, and the industry lobbied the federal and provincial governments for changes in the tax laws. Because the price of gold was fixed, management could not absorb the rising costs by simply passing them on to the purchaser, and notwithstanding the patriotic rhetoric of the mining industry, continued high production levels were not desirable if they were to become less profitable. Rising costs, the increased burden of taxation, and reduced profitability ultimately prompted the mines to seek an acceptable rationale for reducing production. The strike at Kirkland Lake provided the perfect excuse.

The most significant change wrought by the war was in the world market for gold. This change occurred at the same time as the Kirkland Lake strike, and ultimately influenced its outcome. In April 1941, Canada and the United States signed the Hyde Park Agreement to co-ordinate the resources of both countries for the purposes of war production.[29] By November 1941 there was a growing feeling in the American administration that the purchase of gold was an expensive and uneconomical way to finance the Canadian war effort. Accordingly, the United States government ended the priority rating for equipment and machinery used in gold and silver mines outside the United States. The immediate purpose of the American policy was to conserve mining machinery for use in base metal mines; however, in many cases such machinery was indispensable for the continued operation of the existing gold mines. The indirect effect of the American policy was to 'cripple the Canadian gold mining industry.'[30] The cancellation of priorities in the allocation of machinery for Canadian gold mines would mean the slow strangulation of the industry, or at least a deterioration in the net earnings of the mines. As one journalist noted at the time: 'the U.S. is now the supreme arbiter of the present and future health of the gold-mining industry.'[31]

The first response of the Canadian government was to undertake a discreet review of the industry's position to evaluate the degree of Canadian dependence on imported machinery.[32] The government initially hoped to change the American policy. In the face of rumours about the status of the gold-mining industry, a finance department spokesman publicly announced that the Canadian government had no intention of curtailing gold production or placing gold miners in other essential war industries: 'We realize that the production of gold in Canada is necessary for the purpose of stabilizing exchange with the United States and other countries.'[33] Ultimately, the Canadian government recognized that Canadian financial arrangements with the United States had been altered and improved by American purchases of

Canadian goods, and by the Hyde Park Agreement, and concluded that it might 'now become unnecessary and indeed unwise to continue the all-out effort to expand gold production.'[34]

The American entry into the war in December 1941, and the implementation of the lend-lease program, meant that the production of gold was no longer a vital necessity for the balance of payments. Under lend-lease, the method of payment became a minor consideration for goods shipped by the United States to a country opposing the axis powers. Although Canada did not receive direct lend-lease aid, the program indirectly eased the problem of acquiring American dollars, for credit was easily available on American materials falling within the lend-lease categories that were processed or manufactured in Canada on the British account. Gold still retained some importance, but if a large deficit arose again it could be covered more efficiently through lend-lease operations than by gold transfers. For different reasons, both the mine operators and the government concluded that lower gold production was desirable.

During the Kirkland Lake strike, there were recurring rumours that gold production was no longer essential to the war effort and would be cut back. Management desired such cut-backs because of the new burden of wartime taxation, but publicly claimed they were essential because of the effects of unionization. During the strike the union's anxiety about the future of employment induced it to petition the government for clarification of the industry's status: if it were no longer an essential war industry, the gold miners sought to be transferred to more essential works; if it was still an essential industry, the miners sought government intervention to end their strike. The government refused to intervene or answer their questions about the status of the industry. It had realized cut-backs were inevitable, and was only concerned that a downturn in the industry would create staggering economic costs because of the dislocation of many mining communities. It is an interesting sidelight that the government had noted privately that a substantial proportion of the mining population could be transferred to the base metals industry without undue difficulty. This did not occur after the Kirkland Lake strike.

The question of whether or not the gold-mining industry was essential was irrelevant to the main issue in the dispute, but it did affect the outcome of the strike. When the strike eventually occurred, it did not put undue economic pressure on the mining companies, who by that time were prepared to accept a reduction in 'unprofitable' production. There was no urgency to get the strikers back to work, and because of the reduced importance of gold as a medium of exchange, there was no effective pressure for government intervention. At the same time, management blamed the union for the changing circumstances in the industry's ills and for the unemployment that followed the strike. These charges were widely believed, and this reduced the union's effectiveness both during and after the strike.

Management attitudes to the union were a very important factor in explaining the bitterness of the Kirkland Lake strike. The mine operators remained implacably opposed to unions. They never accepted the concept that unrest among the miners was due to real causes, like inadequate wages or poor working conditions. They were profoundly concerned with the fluctuating fortunes of the mining industry and the economic insecurity that this created; but they could not understand that this same insecurity might prompt their employees to organize to protect their own interests. They argued, instead, that the most loyal employees had enlisted and that the war situation had created 'a wonderful opening for the agitator.'[35]

The mine operators were also hostile to the union because they believed, with great conviction, that the best way to run the industry was their way. They feared the challenge a union presented, and resented its competition for the loyalty of 'their' employees. The union was not only a rival source of authority, it represented a repudiation of the values in which the operators believed; its very existence implied a criticism of the way in which the industry was run and the employees were treated.

The myth associated with the development of the mining industry and the mining towns is that extreme individualism contributed to the success of the lonely prospectors who ultimately became mine owners. In Kirkland Lake such men were William Wright, Ed Hargreaves, the Tough brothers, and Harry Oakes. In reality, the development of northern mining, like other resource industries, was a joint public and private enterprise. Governments stimulated and subsidized development, educated personnel in the skills of mining, and contributed through research to the improved technology of the industry. Nevertheless, the 'northern narratives' have repeatedly portrayed 'the self-made, self-sufficient man, driven forward by a sense of destiny, an inspiring self-confidence and an irrational faith in the land.'[36] Hence, S.A. Pain (who was the manager of the Golden Gate mine during the Kirkland Lake strike) described mine owners William Wright and Harry Oakes as two men who arrived in Kirkland Lake independently and 'made themselves millionaires many times over by starting one of the richest goldfields in North America.'[37] He portrayed Oakes as a man who 'knew what he wanted' and 'a tough little megalomaniac ... But you had to be tough at the mining game and *extra tough to be boss of your own gold mine, especially if it was one of the richest in the world*.'[38] When Oakes was prospecting and staking his claims,

Pain suggests that he camped 'in little shelters made of sacking and only had a square meal when he caught a rabbit.' Not surprisingly his dream was to run his own 'one-man' mine.[39]

In such literature, the danger of mining, the luck involved in staking a rich claim, the risks of developing a good vein, and the unpredictability of mining as a business venture were problems that, if overcome, contributed not only to the glory of the few successful miners and prospectors who (with backing) made their fortunes, but served to justify the way they chose to run their businesses. Invariably the strong individualistic mining prospector became an unchallengeable and authoritarian mine owner. After all, these character-istics had contributed to his success. The great mining boom of the 1920s and the 1930s, when all other industries were struggling with the Depres-sion, encouraged the mine owners' confidence in their ability to run the mining business.[40]

The myth of powerful individuals overcoming the vagaries of nature was extremely important because it was believed by the men who ran the mining industry. It affected their perception of their role as owners and employers. As they had built the industry, it followed that they should control their property and run their businesses unilaterally. They were paternalistic; their employees should recognize that what management was doing was for the good of the industry and of the workers. Employees should, therefore, be loyal to their benevolent employers, by whom they were well treated and gainfully employed. Any problems the workers had should be brought to an always sympathetic management. Employee unrest was illegitimate, incon-siderate, and ultimately irrational.

The myth supported such management beliefs and was in conflict both with the increasingly impersonal state of employer/employee relations and with the reality of collective bargaining. The mine managers' perception of the structure of the industry has been called a 'unitary framework' by British industrial sociologist Allan Fox. In such a unitary system, management is one source of legitimate authority, and the only legitimate focus of employee loyalty is the company. Each stratum in the company 'accepts his place and his function gladly, following the leadership of the one so appointed. There are no oppositionary groups or factions and therefore no rival leaders within the team. Nor are there any outside it; the team stands alone, its members owing allegiance to their own leaders but to no others ... Morale and success are closely connected and rest heavily upon personal relationships.'[41] Equally unthinkable is the possibility that substantial sections of the work force could be alienated, or could honestly question management prerogatives, or could demand a role in shaping their own destiny.

To managements holding this view, the right to exercise managerial authority unilaterally is the natural order of things – a self-evident truth that need not be rationally defended. Any apparent discontent with this scheme of things is due to 'stupidity, or short-sightedness, or outdated class rancour, or an inability to grasp the basic principles of economics, or the activities of agitators who create mischief out of nothing.'

By way of comparison, a 'pluralistic' managerial perspective sees business as a plural society containing many related but separate interests that must be maintained in some kind of equilibrium. It accepts the legitimacy of these interests as well as the existence of rival sources of leadership and focuses of loyalty. 'The problem of government of a plural society is not to unify, integrate or liquidate sectional groups and their overriding corporate existence, but to control and balance the activities of constituent groups so as to provide for the maximum degree of freedom of association and action for sectional and group purposes consistent with the general interest of the society as conceived ... by those responsible for its government.'[42] One framework is authoritarian; the other is democratic. The unitary framework does not admit the existence of legitimate sectional interests and cannot accommodate either recognition or negotiation with these competing interests.

This analysis is germane to the behaviour of the mine operators before and during the Kirkland Lake strike. It explains in part why they were so violently opposed to the recognition of the miner's union. As management wrote in a pamphlet published for the public and its shareholders just before the strike: 'Erection of hazards such as CIO inspired labour unrest cannot but strike a blow at the incentive founded on courage, initiative and engineering skill, which produces new found wealth.'[43] Mine operators believed that their managerial prerogative was absolute; the workers might well form a union but management had the right to choose not to negotiate with it: 'It is our privilege as employers ... to decide whether we will negotiate with or consider as bargaining agents for our employees any union which has used dishonourable tactics and ... shown complete irresponsibility with respect to its own word.'[44]

It was not only, as management stated repeatedly, that the CIO was communist-inspired, violent, opportunistic, verging on illegality, and a creator of instability – though these were the reasons given to the public and the government for refusing to recognize the union. Even more important was that the union was 'an outside organization'[45] that did not represent the real interests of the employees. From the management point of view, any unrest in the industry was not the result of genuine dissatisfaction on the part of the miners. Management had operated the mines in the interests of everyone;

there was consequently no real cause for dissatisfaction. The mine operators repeated often that 'employer-employee relations were most harmonious until the CIO started [its] activities in the camp.' They stressed publicly that 'the miners of Kirkland Lake *were* loyal, hardworking and honest citizens.' They pointed out that 'the mines had worked out plans for various kinds of benefits for *their* miners,'[46] including pension plans, insurance group schemes, savings plans, and medical aid programs that had been in existence 'long before the advent of any union organizer.'

What, in the minds of mine operators, was the cause of the recent labour troubles and unrest? For them, it had nothing to do with genuine dissatisfaction on the part of their employees. Rather, the union, which at first had been unsuccessful in 'dividing' their employees' loyalty, had taken advantage of the war situation. Wartime conditions were conducive to union organization; the government had publicly ordered the mines to speed up gold production to obtain foreign exchange; 1500 men had enlisted for war service, and a labour shortage was thereby created. The result was that after twenty years 'of excellent relations between mines and miners [the union] sensed an opportunity to capitalize on a well-paid community and entrench itself for an attack on Canada's vital base metal mines.'[47] Predictably, from the mine operators' point of view, such 'trouble' caused some men to move to other camps, efficiency dropped, and production declined.

Workers had joined the 'so-called union,' not because they had different interests and wanted more money or security, but because they had been *tricked* by professional organizers who told lies and 'utilized hopeless and illegal promises to attract unthinking followers, and employed threats and coercion to compel men to join the union.'[48]

The mine operators concluded that the fight for union recognition was a selfish desire on the part of organizers to guarantee the 'flow of funds into the union coffers.' This 'selfish motive' was more important than the welfare of the employees 'in the minds of the CIO professional organizers and negotiators.' Consequently, management, who had always had the employees' welfare at heart (and particularly the interest of those miners overseas, whom, management was sure, if given a voice, would have stoutly rejected any union out of patriotism), refused for the employees' own good to recognize the union and to allow it to 'take over' the industry. Management's apparent belief that the union *wanted* to control the industry and entirely supplant its authority demonstrated its real fear of the changes that the introduction of collective bargaining would bring to the mining industry.

The mine operators refused to recognize the union, but did agree to bargain collectively with their own employees through committees chosen by the miners. The union saw this position as 'company unionism' – a device frequently used in the United States to undermine independent employee organizations, and that was, by this time, illegal in the USA. Management believed that wages, hours, and working conditions were fair. It asserted that these conditions had not been questioned by the workers, because the workers themselves had not come to management personally to complain. In its view, the record 'reveals no lack of willingness by the mines to enter into any understanding that will contribute to the maintenance of conditions beneficial to the employees.'[49] The mine operators completely rejected the idea that the organization of the union was a sign of employee dissatisfaction. Past policy had made for success in the industry and harmony between the workers and management. There was no need for change. Even after the miners had rejected the idea of employees' committees, the mine operators continued to try to establish such committees as an alternative to the union.

The committees were acceptable to management because, in practice, management would dominate them as it did in the United States. So long as employees could not make common cause with workers in other organizations, their bargaining power would be reduced. As long as management retained the power to demote, lay off, transfer, or fire 'agitators' or 'disloyal' employees, and reward 'loyal' employees, the committee and its leadership could be effectively manipulated. There was no possibility, given management attitudes, that a committee would be truly independent. The committee solution was viewed by management as a way to curb industrial unrest among its employees, while at the same time preserving the mine operators' exclusive managerial prerogatives. The underlying assumption remained the same: only management had the expertise to decide what was good for the company. The independence, unpredictability, and possible opposition of a trade union was intolerable.

The mine operators' perception of the reasons for their employees' support of a union and the resulting strike was distorted, but it was firmly rooted in the past history of the industry. It meant their position would be especially obdurate throughout the strike. The other more practical economic reasons for resisting the union were important to the mine operators, but their 'unitary frame of reference' explains the intensity of their resistance to the very recognition of the union. Existing government wage controls greatly reduced the potential economic impact of collective bargaining. In this respect the management perception was accurate, for it was not profit, but unfettered managerial authority, that was threatened.

This management position refused to recognize a number of important structural changes in the gold-mining industry that had affected employee interests. For mine employees, changing economic circumstances had eroded the force of the myth that individual initiative brought wealth and prosperity to all. Employees no longer believed that the judgment of the mine owners was unchallengeable or that management always had the workers' interests at heart. The miners sought a voice in matters directly affecting their employment conditions and chose a union to make their case. They had no intention of taking over the industry as the mine operators charged, but they did seek greater participation and consultation within the decision-making structure. To management this was intolerable. This was the basis for the conflict in the Kirkland Lake strike, as it was in many other recognition strikes undertaken by CCL unions in these war years.

THE CHARACTER AND CONDITIONS OF THE WORK FORCE IN THE MINING COMMUNITY

General economic conditions, changes in the size and technology of the mining industry, and special characteristics peculiar to mining communities and their work forces were all important factors contributing to the unionization of the Kirkland Lake miners.

Historically, miners have exhibited a high propensity to join unions and to strike. Kirkland Lake is just one example of this pattern. Miners live in geographically isolated communities that are frequently dominated by a single industry that profoundly shapes the character of the town. The lives of almost all employees will be influenced by the same events and, in particular, by the operation of the dominant industry. Industrial hazards, fluctuations in employment, low wages, and poor housing are commonly cited as employee grievances. It is important that, in a one-industry town, virtually all employees will have the *same grievances* at the same time against the same people. The principal employer is often the landlord, grocer, and policeman as well. As a result, anger and frustration are accumulated and concentrated against a single target.[50] This situation is exacerbated if control of the industry itself is removed from the local scene. Ultimate control of company policy did not rest with the management in Kirkland Lake, but with the boards of directors whose members lived in Toronto, Montreal, New York, and Chicago: 'Those members are presumably not in close touch with local conditions but their decisions set limits to the freedom of local management.'[51] Consequently, local management, despite its greater ability to judge local conditions than that of the absentee directors, was unable to deal with them independently. This situation increased the employees' sense of isolation. The combination of arrogant paternalism by the local managers and absentee ownership contributed to distrustful employer/

employee relations. Management's rhetoric about past personal relations and its present willingness to hear individual grievances seemed meaningless to the employees.

Miners are a homogeneous group with shared grievances and problems, but in addition their skills are not easily transferable. Mobility between industries is difficult. Protest is, therefore, likely to result in moving to another mining town or to a mass walkout. In the case of the Kirkland Lake miners, both traditional outlets were used. Before the strike, miners moved to other mining towns. When the strike was lost, many miners were forced to leave and find work in the manufacturing sector – a situation that was only made possible by the willingness of industry and the government to retrain workers for war work.[52]

Miners have little or no social mobility. In Kirkland Lake, 'you were a mine owner or you were a worker.'[53] This social stratification accentuated the divisions within the town and caused the strike to take on political characteristics. In such circumstances a union can become an important vehicle for change; a strike tends to be partly a substitute for occupational and social mobility. In Kirkland Lake the union took on the characteristics of a working-class party that was mobilizing employee interests against those of their employers. The physical isolation and the social stratification of Kirkland Lake created a community of interest among the miners and made the strike a form of class conflict with the employers in the community.

Finally, the nature of the job tends to encourage militancy: 'If the job is physically difficult, and unpleasant, unskilled or semi-skilled, and casual or seasonal and fosters an independent spirit ... it will draw tough, inconstant, combative and virile workers and they will be inclined to strike ... Sailors, longshoremen, miners and lumberjacks are popularly accepted as being more vigorous and combative types ...'[54] The mining industry had attracted a certain kind of worker. During the Great Depression, when the industry was among the few hiring, it had a great pool of labour from which to choose. Consequently the miners chosen were 'the cream of the crop'[55] – men chosen out of thousands of unemployed. In the Depression 'some of the best workers in the land were drifting and travelling on the lookout for some means of earning a living once more ... They started coming here [to the North] in the hundreds.'[56] As one Kirkland Lake resident noted during the strike: 'The Kirkland Lake camp gets more than its share of strong, intelligent, healthy, ambitious young men from the farming and industrial districts of Canada.'[57]

Long-term job security remained a principal concern, but by 1940 the mines were operating at peak production and the employees believed (wrongly as it turned out) that it was less likely that they would be laid off if they engaged in union activity. Many of the younger miners had a sense of security because there were now alternative job opportunities in Southern Ontario, and the work force in the mining industry was relatively experienced and, therefore, hard to replace. These factors all contributed to unionization of the miners and gave the men a sense of confidence during the strike, a sense of loyalty to the union, and a feeling of defiance towards an increasingly intransigent employer group.

The nature of the gold-mining industry had changed from its beginnings in a way that profoundly affected the position of the mine workers in the industry. The young adventurers of the 1920s, expecting to strike it rich in the gold fields with their own claims, were largely a part of the past. The young workers in the 1930s, expecting to at least find work at relatively good wages in the midst of a depression, had, by the 1940s, become married men with families and increased expenses. The results of a 1941 union survey of 1000 employees working in the Kirkland Lake mines dispelled the romantic notion of the independent, transient miner.[58] It found that only 9.5 per cent of the miners were twenty-five years of age and younger. Most of the men (62 per cent) were between the ages of thirty and forty-five. Of the 1000 surveyed, 83.6 per cent were married. The results demonstrated that the men employed in the mines were a stable group of workers, that 35.6 per cent had been employed in the Kirkland Lake area from ten to fifteen years and 67 per cent for more than five years, and that only a small percentage (9.9 per cent) were recent employees (with the companies for less than a year).[59]

The cost of living was rising during the war, and necessities were always more expensive in the northern mining towns than in the metropolitan centres. The miners found that they had to pay 20–50 per cent higher prices for clothing and food than those living in the farm communities farther south, and twice as much for rent. The business of mining incurred further costs such as that of special clothing, which was expensive and wore out quickly from the rough work.

During the strike the mining companies pointed to the prevalence of consumer goods as proof of their adequate wages, but there is evidence that advertising was merely 'encouraging these poor workers to enjoy the good things of life on the time payment plan.'[60] Miners had small pay cheques and long lists of obligations. Any extras were bought on credit, which was readily extended by the merchants of the town. There is also evidence, however, that life was not easy for creditors in Kirkland Lake. The miners were frequently in default. In one year there were nearly twice as many divisional court actions for debt issued in the communities of Kirkland Lake and Timmins than in the City of Toronto. The provincial government found it necessary to establish a new divisional court in Larder Lake in order to cope with the increase in the number of court actions. One collector in one of the many new collecting agencies in Kirkland Lake believed the average miner in the camp was usually forced to live 'too close to the line' and that 'the rate of remuneration for his services is too low in proportion to the high cost of living in the Northern camp especially so when the hazardous, debilitating nature of his working conditions is taken into account.'[61] In this situation good health and consistent work were vitally important. Mining was conducive to neither. Debtor miners with initiative, who found they were worse off than when they started to work in the mines, often would simply move to a new camp to make a fresh start.

It is not surprising that workers were attracted to the union as a vehicle to give them some collective strength. The union developed in response to changes in the industry that had an economic foundation and, as a result, unionization was not a temporary manifestation.

It was clear to all but the mine owners and their supporters that the method of running the industry had to change to better accommodate the needs of the miners. Increases in taxes were grudgingly accepted, but changes desired by their employees were out of the question. The owners were insensitive to the workers' needs, having become separated from their work force. A medical plan and employees' committees were introduced too late to effectively forestall the organization of Local 240.

The increasing economic difficulties of the miners contrasted with well-publicized accounts of the profits of the gold-mining industry. The workers decided that the time had come for them to have a greater share in the wealth. A union pamphlet on the Kirkland Lake dispute stressed this view and noted that the twelve mines had produced $25 million a year. In the previous ten years, during the 1930s, costs had declined, but the mines had had an 85-per-cent increase in the value of gold and a guaranteed market. In five years the miners had received a five-cent-an-hour increase in wages.[62]

The successful organization of a trade union and the prosecution of a strike require a high degree of group cohesion and solidarity. Racial, ethnic, or religious differences within the work group usually impede unionization. According to the 1931 census there were 9915 people in Kirkland Lake, of whom 50.7 per cent were of British origin and 13.83 per cent of French-Canadian origin. Canada's two founding peoples made up approximately 65 per cent of the town's population. The next largest groups were Finns (7.2 per cent), Poles (4.2 per cent), Scandinavians (2.9 per cent), Ukrainians (2.7 per cent), and Italians, Germans, Jews, and Romanians each with less than 2 per cent. Groups smaller than this were counted together and totalled 12.32 per cent of the population. Between 1931 and 1941 the numbers of inhabitants in Kirkland Lake increased rapidly, but the population mix remained similar. More immigrants were employed in the mines, but not as many as the comments of the mining companies during the strike suggested. In 1941 the population of Teck Township[63] had risen to 20 409, of which 14 873 were derived from Canada's founding peoples. The rest were of European origin – the largest groups being northern Europeans (Finns and Scandinavians, etc.), followed by eastern European groups like the Poles. This racial preponderance was true of Cobalt, Larder Lake, and New Liskeard; while Kirkland Lake was not listed specifically, it is a reasonable assumption that the composition of its population followed the same pattern. Once the war began, 1500 predominantly British-Canadian miners enlisted in the armed forces and they were for the most part replaced with immigrants. There is evidence that the middle-class citizens in the community particularly resented what they considered an influx of 'foreigners' and feared the changes that this presence might bring to the town.

According to the breakdowns of ethnic origin of the Kirkland Lake mining work force, the gold-mining companies, with the exception of two, employed a larger proportion of Canadian and British workers[64] than might be expected from the ethnic distribution of Kirkland Lake's population. An overall average of all the companies (30 November 1940) showed that they employed 73 per cent Canadian- and British-born workers and 27 per cent foreign-born workers. Of the foreign-born, more than half were naturalized Canadian citizens (except in three mines).[65] On an overall average, 52 per cent of the foreign born were naturalized.

Ethnic differences within the work force did not seriously impede union organization, perhaps because of the open employer hostility to all of the 'foreign' elements. In any event, the local was predominantly Anglo-Saxon in most cases, as were the majority of miners. As in other labour struggles in the history of the Canadian labour movement, the solid trade union tradition of British-born workers was a positive advantage to the union. The tolerant social attitudes of the union and the sense of solidarity among the miners overcame whatever racial tensions there may have been. During the strike the operators announced on one occasion that 43 per cent of the men still on strike were 'of foreign extraction.' The union protested to justice minister St Laurent and charged that such announcements were 'a calculated attempt to create racial bitterness.' The union alleged that the operators could be prosecuted under the Defence of Canada Regulations, since inciting prejudice affected the efficient prosecution of the war.[66] On another occasion a union radio broadcast condemned the appeal to racial hatreds that the mine operators had made and asserted that 'Local 240 makes no distinction because of race, creed or colour. It is proud of its membership – a fraternity of free men regardless of their racial origin, their religious beliefs, or political associations.'[67]

To some extent the past work experience of the foreign-born miners may have benefitted the union. The local was charged with being communist controlled prior to and throughout the strike. This anti-union propaganda by the companies and the press was a tactic that was largely ineffective. The foreign-born workers were not influenced by such charges. Many of them – Slavs, Ukrainians, and others – had worked under factory bosses in Europe, had developed a class perspective that was not altered by the environment of Kirkland Lake. 'They knew the score'[68] in a polarized strike situation, and stuck with the union.

In addition to the young men who had travelled north during the Depression to find work, there was another group of workers who had lived in the North for many years. They were the ones most likely to remain after the strike and consequently they felt they had the most to lose by the strike. Yet their expectations of what a strike would achieve were influenced by the past history of the mining industry. Many of them remembered the Cobalt silver bonus of 1918 paid to miners to help offset World War I inflation, and rescinded after the war. Their fathers had been in the silver mines at that time. When Local 240 drafted its strike demands, the bonus principle once again had become wartime economic policy. These workers wanted higher bonuses but worried about the insecurity associated with this type of payment in comparison with a basic rate increase. These 'Northerners' remembered what had happened to their fathers and supported the union.[69]

For all these reasons – the nature of the miners' work, the character of the persons attracted to such work, the changing economic situation, the ethnic composition of the work force, the stirrings of the war economy, and the general insecurity – the workers decided that they needed a union.

THE EARLY HISTORY OF MINE MILL IN NORTHERN ONTARIO

Rugged individualism and single-minded determination were not characteristics exclusive to the prospectors and the mine operators who developed the mining industry. Such qualities were exhibited by many miners as well. Miners like John L. Lewis were prominent in the leadership of the CIO movement in the United States and, to a lesser degree, in Canada.[70] Historically, miners have been highly unionized and the Kirkland Lake miners were no exception. Once the miners in Kirkland Lake decided to unionize (and it was a long process), they became united and very committed to their union.[71]

The International Union of Mine Mill and Smelter Workers (Mine Mill) and its predecessor, the Western Federation of Miners (WFM), have had a dramatic and colourful history. The miners in Mine Mill believed that their struggle was a continuation of the one initiated by the WFM in 1893[72] and interpreted events in the light of the union's heritage.

The WFM was formed after a bitter strike at Coeur d'Alenes, Idaho, in 1892, and at one point had 40 000 members in 200 local unions.[73] Management opposed the 1892 strike by importing strike-breakers, and in the resulting conflict the strike was broken by the combined intervention of strike-breakers and police in a pattern that would recur often.[74]

Unionism in the non-ferrous metals industry grew slowly, but eventually organization was extended to the miners in Butte, Montana, and to metal mineworkers in Utah, Alabama, Oklahoma, Missouri, and Kansas.[75] In Canada, the WFM organized primarily in the Canadian West – particularly in British Columbia – but it also had a sprinkling of members as far east as Northern Ontario.[76]

The WFM in the United States faded after two decades of struggle against extreme employer opposition. In 1903–04, 'labour wars' in Colorado involved employer-financed military intervention, the break-up of a labour press, false arrests, threats, disregard of court orders, and the exportation of innocent workers from the scene in box cars. 'Mine Mill's attitude toward police intervention – such as that of Hepburn's Hussars in the Kirkland Lake strike of 1941–42 – was developed from such accumulated experiences as those under men like Adjutant General Bell [in Colorado].'[77] Despite these long strikes, which sapped the union's strength, it never achieved recognition.

The WFM was further undermined by weak leadership, an inability to sustain vigorous organizing among the miners, and internal disunity. The latter problem originated when the WFM, an industrial union, refused to affiliate with the craft-dominated American Federation of Labor (AFL) because it considered that organization to be 'effete, misguided, and improperly conceived and led.'[78] Instead it brought together other unions to form the Western Labor Union. This was later extended east to become the American Labor Union. These organizations did not flourish, and in 1905 the WFM sponsored a conference in Chicago from which was born the Industrial Workers of the World (IWW) – a militant, radical syndicalist, industrial union, intended to displace the American Federation of Labor.[79] Dissension between rival IWW and AFL factions within the WFM led to internal union wrangling and to AFL efforts to set up rival unions within the WFM jurisdiction. Employer opposition to all miners' unions resulted in the loss of membership. Belated affiliation of the miners' union to the AFL did not eliminate factionalism within the union.

During World War I there were temporary membership gains, but the destruction of the miners' union in the large local in Butte, Montana, in 1914 finally broke the back of the WFM. The final dissolution of the union followed the post-war depression and a stagnant period in non-ferrous metals' production, as when International Nickel in Sudbury in 1921 shut down and threw two-thirds of the labour force out of work.[80] In 1916 the WFM had been renamed the International Union of Mine, Mill and Smelter Workers (IUMMSW), but the union remained moribund until it was revived by New Deal legislation in the 1930s.

The WFM first entered Canada on the west coast, establishing a local at Rossland, BC. In 1899, the union established District no. 6, which by 1901 had eighteen locals in British Columbia. In 1903, the United Brotherhood of Railway Employees (UBRE), an affiliate of the American Labour Union, struck the CPR, successfully tying up freight traffic between Vancouver and Winnipeg. Coal miners, organized by the WFM and provoked by mine operators' threats, struck in sympathy. A royal commission, established to investigate the situation, included the then deputy minister of labour, Mackenzie King. The commission's report, substantially written by King, condemned the union's actions on the grounds that the international and radical political character of the UBRE and WFM imperilled the sovereignty of Canada.[81] The strike was a failure, but the union's activities laid the groundwork for the radical orientation of the Western labour movement in the 1920s and 1930s.

The WFM locals in Northern Ontario were organized a decade after those in British Columbia. An effort to organize a local in Sudbury in 1905 had failed. The first success was Local 146 – chartered in Cobalt in 1906. The WFM grew steadily until 1912 when it reached its maximum Canadian membership, which was equally divided between British Columbia and Ontario. In 1914 the WFM had twenty branches, including Cobalt, Kirkland Lake, Sudbury, Gowganda, and Porcupine in Ontario; and Nelson, Rossland, and Trail in British Columbia. Membership in Canada at this time was estimated at 4000, and the total union membership at 65000. Between 1914 and 1919 other locals were established in Quebec and Ontario, but these failed. In Eastern Canada, Ontario members were included in District no. 17. These locals could not sustain themselves at a time when the international union was declining.

At the end of the war, inflation contributed to unrest among Canadian miners, including those in Northern Ontario. In these circumstances, when both the union and the industry were in their infancy, the miners in Kirkland Lake struck. This strike was an important test of strength for the organization, although the Kirkland Lake miners had not expected to initiate strike action. They had expected a strike in neighbouring Cobalt where miners were equally discontented.[82] The silver miners in Cobalt had received a higher silver bonus during the war, but there was no advance in wage rates or in real income as a result of wartime inflation. This situation created great insecurity, particularly when the price of silver fell following the war. There were other issues in the Cobalt dispute apart from wages. The union had been active in the camp for years, but had never been able to negotiate with management. Working conditions, better and cheaper housing, a health care plan covering compensation for accidents, and hospital referrals to the city for sick and injured workmen were all concerns of both the union and the community.

Underlying these specific issues was the general issue of employees' control of their working and living environment. In 1919 this was the root cause of dissatisfaction in Cobalt. Expropriation was a real possibility. This created great insecurity, as did the unstable production patterns in the industry. Such problems were 'endemic to the very nature of the mining camps. To a great extent such towns were in existence only because the mines and the companies were in business.'[83] In the newer camps like Kirkland Lake, the mine owners owned all the land and leased property to workers on which they could build homes. Despite management's publicly expressed concern for the miners, the workers were seen as a factor in production to be utilized in accordance with the demands of the market. The threat of unemployment was always in the background. Management could be generous to individual workers but was adamantly opposed to their organization. In the miners' view, management was autocratic and paternalistic. These same grievances emerged two decades later in the Kirkland Lake strike of 1941, and the tactical problems in the earlier strike influenced the strategy adopted in the later one.

On 27 May 1919, the Kirkland Lake miners' local presented their demands to the gold-mining operators, and on 2 June Cobalt did the same with the silver-mine operators. The most important demands were minimum wages, an eight-hour day, forty-four-hour work week, union recognition, collective bargaining, and a dues check-off for the union. The *Northern Miner* reported that the mine operators would consider neither recognizing the union nor raising wages. Because of the silver bonus, the Cobalt miners were still earning more than the gold miners in Kirkland Lake; consequently, the *Northern Miner* predicted that a strike was inevitable, and that it would begin in Kirkland Lake.

All the local unions in the mining camps were affiliated with Mine Mill. Cobalt was the national headquarters of the union's District no. 17,[84] which called a meeting to seek support for the Cobalt local. The meeting discussed the possibility of a general strike. The Kirkland Lake local pressed for a sympathetic strike in order to tie up the entire industry. The Gowganda and Porcupine locals accepted the mine operators' arguments about their inability to pay higher wages due to high costs.[85] They advised against a strike, but urged financial and moral support for the strikers in Cobalt. On 8 June the Cobalt miners took a strike vote but they later met with Senator Gideon Robertson, the minister of labour, and deferred strike action. 'Kirkland Lake, to breathe life into its position, feeling that it was very, very right'[86] walked out on 12 June, supposedly in sympathy with the Cobalt miners, but actually before the Cobalt local had struck. The strike in Cobalt was further delayed when Department of Labour officials arrived in Cobalt on 14 June. The silver miners were hoping that the forthcoming Industrial Relations Commission Report (1919) would favour union recognition. It did, but that did not affect the immediate strike situation in the Northern gold and silver mines.

The timing of the Kirkland Lake strike action created considerable confusion. The Kirkland Lake local had struck without the approval of the district, and Cobalt workers were unsympathetic to this unilateral action.[87] However, once on strike, the employees could not go back to work. The effect was to undermine the unity and strength of organized labour in the Northern Ontario mining camps. As a result, the strike activity never expanded and the government was never forced to intervene or bring about a settlement.

In the Kirkland Lake strike of 1919, 525 miners remained on strike from 12 June to mid autumn. The strike resulted in 56 176 lost man-days[88] and a general exodus of miners from the camp. As the hardships increased, those who were mobile drifted to other camps. The Cobalt strikers were more settled and less able to move.

In 1919 the Kirkland Lake local had no money. It received donations from other locals – but not enough. Much of the money in the Cobalt local, set aside for its strike, 'had to be used to feed the miners of Kirkland Lake.'[89] Consequently, when the Cobalt miners struck they were in a weaker financial position.

Both strikes were lost. The Cobalt local terminated its strike in September; although the Kirkland Lake strike continued somewhat longer, nothing was achieved for the workers of either local. In Cobalt, the workers were permitted to elect a committee composed of representatives of *individual* mines that could negotiate with the mine operators in place of a union. The workers' own organization had been defeated and the new arrangement was organized in such a way as to ensure that the employee representatives were subject to the pressures of the mine managers. The Cobalt 'employee committee' was a symbol of the union's defeat, but it also demonstrated the futility of accepting an impotent employee committee as an alternative to independent trade unionism.

The disunity among the miners' locals had proved to be a great weakness, just as it did in 1939 when the miners' local at the Teck-Hughes mine bargained independently and failed to get a collective agreement. Gradually it became evident that the gold miners at Kirkland Lake must organize into one local and demand a master agreement to cover all the mines. All the gold-mining companies in the area had to be treated as one employer for collective-bargaining purposes. All the miners had to strike at the same time. The 1941 strike was fought and lost by a *unified* group.

The Kirkland Lake local was caught up in the mood of militancy of 1919 (it called its executive a 'soviet'); the atmosphere of that crisis year in labour history influenced the local's strategy of a general walkout and determined its precipitate action. Had there been general support from all the miners' locals, the workers might have succeeded. As it was, the union fared badly in both strikes and the twin defeats 'practically closed out the Ontario district'[90] for several years.

The fluctuating fortunes of Mine Mill's organizing drives in the northern mining industry caused its supporters grave concern. After the 1919 strike, 'there was realization within the framework or within the minds of the members of Mine Mill that there is something wrong with this union. And many times it would flare up, build up a treasury, build up a great membership and it went on a great strike and it would deplete its treasury and it got nowhere.'[91] In their disillusionment, half of the membership responded to the appeals of organizer Joe Knight who was touring the area, and for a brief time they became members of the One Big Union (OBU). It was a choice between the failing leadership of Mine Mill, which had provided little assistance in the recent strike, or the rising star of the OBU[92]: 'But the OBU wasn't qualified for it. They weren't miners that organized and they didn't speak the miners' language – they were mostly railway men and clerks ... that come out from Winnipeg.'[93] They were good orators, but once they left, they left nothing behind them. In these mining camps the OBU developed along lines similar to the IWW. They set up no committees. Their romantic belief in the natural solidarity of the labouring masses and spontaneous collective action did not require continuous organization or envisage the servicing of locals over a long-term period. This was an approach that the OBU shared with the IWW[94] and, as a result, it achieved neither revolutionary change, nor 'plain and simple' collective bargaining. It did not fulfil the workers' real need, which in the case of the Northern Ontario miners was for a strong industrial union with sufficient bargaining power to protect workers from authoritarian management and the fluctuations of the market. The result was that the Kirkland Lake OBU local broke down and left nothing to build upon except memories and ideals.

Throughout the 1920s and early 1930s there were only a few weak locals in the IUMMSW. After 1925, Mine Mill ceased to function in Canada.[95] By 1932 not more than six locals were in existence in all of North America. It was not until the New Deal, the passage of the NIRA, and the birth of the CIO encouraged the growth of many other new industrial unions that Mine Mill could be revived and developed again in Northern Ontario.

Between 1919 and 1936 there was little successful union organizing in Canada. In the mid 1930s the Mine Workers' Union of Canada (MWUC) (affiliated to the Workers' Unity League [WUL]) made a feeble effort to organize in Northern Ontario, but in 1934 the Sudbury local of the MWUC dissolved and its locals in Timmins and Kirkland Lake broke up in 1935. The rebirth of Mine Mill in Northern Ontario began in 1936. Mine Mill representative Tom McGuire remembers that 'at this time, throughout most of North America, there was a tremendous need for organization in the United States and in Canada. Workers everywhere wanted some medium of expression and security.'[96] Before 1936 there was a need for union organization, and after 1936 there was a demand for it: 'Men in many of the mining camps and smelter towns were ripe for unionism. Depression privations had not cowed them to the point where self-help did not beckon, and they wanted a voice in determining the conditions of their working lives.'[97] The much publicized profits of the gold mining industry had reached unprecedented levels, and the miners wanted their share of the wealth. Many young men had entered the mines during the Depression and they were profoundly dissatisfied with existing conditions. The example of American workers organizing under the CIO banner, and of the auto workers organizing in Oshawa, encouraged a desire for their own union: 'We all moved about the same tme in the "hungry thirties."'[98]

The building of the Northern locals occurred between 1936 and 1941, and the growth of these locals parallelled that of the international union. In 1936, Reid Robinson, a 'new unionist' and president of the Butte Miners'

Union, became president of the IUMMSW. In that year Mine Mill was still a weak organization, having recognition in only a few widely scattered places throughout the industry. It had a long way to go before it could bring unionism to all the important ore and metal producers and fabricators.[99] The new leadership was intent on accomplishing this goal.

In 1936, organizing in Northern Ontario was directed from Sudbury by George 'Scotty' Anderson – the sole organizer on the international payroll. The union campaign started with an attempt to organize a local at Inco, since it was believed that a victory there would facilitate organization in the other mining camps in the province.[100] The union recognized the significance and power of Inco in the industry. It also understood that organizing in Sudbury would be very difficult since it was reputed to be a centre for 'scabs.' Nevertheless, a group of people met in Sudbury to plan the return of the union organization to the mines of Kirkland Lake, Timmins, and Sudbury. The group consisted of people from Kirkland Lake, a few old OBU members, Canadian mine workers from the defunct MWUC, and Timmins and Sudbury people from the remnants of the WUL, which had dissolved itself in 1935 on orders from the Communist International. These groups applied for three charters from the IUMMSW. Sudbury received its charter as Local 239 in April 1936. By May it had recruited 150 members after a tough campaign in which the organizer was molested by thugs and the union office was wrecked. In May 1936, Local 240 in Kirkland Lake was chartered and in June 1936, the Timmins local was chartered as Local 241.[101]

The original aim had been to organize in Sudbury but the union did not, in fact, link its development in Northern Ontario solely to the organization of Inco. There were local organizers in all the Northern Ontario mining towns. The union's organizing drives benefitted from the influx of experienced communist organizers after the WUL was disbanded. The miners' union had the solid support of the Communist Party of Canada as well as other workers' organizations; locally the Ukrainian Labour Farm Temple Association (ULFTA) played a particularly important organizing role in 1937.[102] The union published a monthly bulletin for the Northern Ontario locals called *Union News*. Despite difficulties, there was an immediate response to the union's campaign; in 1936, 2000 copies of the paper were being sold. By 1937 Anderson believed that shortly the union would be in a position to present its demands to the mine operators, that a solidly supported strike would follow that would begin in Timmins (which was the most organized) and would spread throughout the other towns. The strike never took place.[103]

It was rumoured that the international union and the CIO were willing to spend $1.5 million to organize the mining and smelter workers in Northern Ontario. In November 1937, James Robinson (father of Reid Robinson, the future president of the union) visited Kirkland Lake and then toured Timmins and Sudbury. As an undercover man employed by the Ontario Mining Association reported to the provincial government, Robinson told the Kirkland Lake local executive that

he had been sent by the C.I.O. to look the field over here as it was decided at the last C.I.O. conference that all industries in Ontario would be organized. He said, 'I am quite sure that after they learn of the size of the mining field here that you can expect plenty of action and support from them ... When I return to Detroit, I will ask for every available organizer and every dollar that can be raised be concentrated right here in Ontario. Your Premier Hepburn has challenged the C.I.O.; he will see what he is bucking in the coming year. The members of the C.I.O., even myself, little realized the greatness of the field here for the organizational gains. You can be sure that I will do everything in my power to get quick action here.'[104]

Robinson's expectations of success were not fulfilled, for even in 1937 there appeared to be problems. Robinson warned the Kirkland Lake executive (on which a number of communist sympathizers sat) that the organizers were getting away from the union 'by mixing too much in the politics of the district' and that in future they must stay away from any communist activity. A good organizer had to inspire confidence in those he was trying to organize and must not waste time on matters that did not concern the union. The organizers had worked in a half-hearted way for the union and had discouraged a large number of miners. 'He had talked to the miners from all the Ontario mining districts and he was only repeating what they had said to him, and they were the men who would be the mainstay of the Union and their wishes had to be considered first.'[105] Robinson also warned that any money collected in the form of dues was to be placed at the disposal of the union and not to be used for other activities. Tom Church, a local organizer, and others were given notice that if they wanted to participate in the 'new program,' they had to change their ways. He strongly urged that many union meetings be held to build the union.

The response of the companies to this extensive organizing activity was predictable and effective. In September 1936, both Falconbridge and Inco gave a voluntary increase in wages of 5 per cent. After SWOC signed a contract with Inco in West Virginia in 1937, Inco increased Canadian wages by 10 per cent, and predictably, began to organize a 'company union.'[106]

Despite a relatively favourable response (it was estimated that between 30 per cent and 60 per cent of all the mines' work forces were organized at this time), the new locals were plagued by a shortage of funds. In Sudbury, the local lost its momentum and dissolved in 1938, in part because Anderson had been taken off the union's payroll for lack of money. Falconbridge Local 278, set up in 1937, dissolved a year later. Once again the pattern of temporary success followed by rapid decline seemed to be emerging. In 1938, Mine Mill had 4000 members in Canada. In 1940 it had 176.[107] These marked fluctuations were attributable to lack of funds supporting the organization and to internal disunity.

For some time, the Kirkland Lake local, formed in 1936, remained an underground organization. Most of the charter members were members of the Communist Party: 'At that time, they were the only people that would dare go out and take a chance on this organizing.'[108] The organizing was carried on quietly, and progress was made. About 55 per cent of the work force had joined the union in 1937 and 25 per cent of the non-union members were expected to support a strike if it came to that. The majority of members were foreigners, especially Ukrainians and Finns, although there were some Poles and Hungarians as well. One problem was a split between the foreign workers on the one hand and the English and French workers on the other hand. 'The cause that the union is not organized is the fact that the English workers are against the C.I.O. and ULFTA as chief organizers of the miners' union in this town.'[109] The same split had also occurred in Timmins where Max Kaplin, the owner of a haberdashery and a sympathizer of the union, recounted to an undercover agent that the majority of English workers were opposed to ULFTA members as organizers, although they had no objection to the CIO. He believed that a good number of English workers would join the miners' union if ULFTA dissociated itself from the union. The basis of this split was possibly racial, but also political.[110]

The other great barrier to union organization in Kirkland Lake was employer opposition. Robinson was aware of this on his visit. 'In the Ontario mines ... the mine operators had been able to organize a very efficient spy system, and there is no doubt that they are spending plenty of money to forestall the growth of the Union; this particularly applies here in Kirkland Lake.'[111] The companies retaliated by firing individuals they suspected of union activity and blacklisting them from the industry. At the Lake Shore mine, 270 men were fired because a single union application card was found on one level of the mine: 'They took no chances even including stoke bosses ... that's typical of what went on in most of the mines – along with watching you closely.'[112] Despite difficulties in organizing in the face of such employer opposition, the local was quite strong by the summer of 1940 and the union's resources were increasingly concentrated where the Union had had the most success – in Kirkland Lake.

3

The Teck–Hughes dispute and its aftermath (October 1939–July 1941)

The dispute at the Teck-Hughes mine in 1939–40 was the precursor of the later and larger confrontation in Kirkland Lake in 1941. The central issues were the same: union recognition and the inadequacy of the government's conciliation policy in circumstances in which the employer refused to negotiate. The experience at Teck-Hughes influenced the union's strategy in 1941, and helps to explain the miners' later suspicion and distrust of the government.

The organizing campaign at the Teck-Hughes mine was typical of those being conducted by Mine Mill in the other mines, except that it was more successful. The membership drive began with the circulation of two leaflets.[1] A third leaflet explained the procedure that the union intended to follow in order to bring about negotiations. On 3 October 1939 the union sent a letter inviting the owners of the mine to meet with union representatives. The mine owners did not reply by 14 October as the union had requested; consequently a meeting was held the next day. Some 150 to 175 employees (or about a quarter of the mine's work force) attended, endorsed the union's organizing campaign, and issued 'lists of authorization' to be signed by members. Actual soliciting of additional members did not begin for another week. The union issued another leaflet explaining its campaign, and held another meeting. By mid November, the union's lawyer, J.L. Cohen, advised that since the employer refused to meet them, it was appropriate to apply for a conciliation board.[2]

Under existing federal legislation, a strike was the only way that a union could force a reluctant employer to bargain. Conciliation was a compulsory prerequisite to a legal strike. Both parties were required to present their positions to the conciliation board, but they were not compelled to meet or to bargain directly with each other. This procedure was inadequate in cases where union recognition was being refused by an intransigent employer. The legislation accorded status to the union and required it to appear before the board as the employee's representative before those employees were allowed to strike; however, nothing required the employer to recognize the status of the union, or to entertain its submissions. Conciliation was a mechanism for avoiding strikes; it neither encouraged nor supported collective bargaining. Indeed, conciliation standing alone may have undermined collective bargaining. There was no requirement and no incentive to meet the union to exchange submissions prior to conciliation. In any event, the recommendations of the board were in no way binding. The employer could implement them in whole or in part or ignore them as he saw fit, and the delay gave him time to retaliate against staunch union supporters and to prepare for a strike.

The union at Teck-Hughes had no other alternative. Cohen explained that the correct procedure for requesting a conciliation board inquiry was to conduct a vote at a union meeting or circulate a petition authorizing several individuals to apply for a board. He suggested that the union make very *specific* demands because these would establish the scope and terms of reference for the board's inquiry. The union followed this advice and made demands similar to those they would make in 1941–42 – a fifteen-cent-an-hour wage increase to meet the rise in the cost of living, and the recognition of Local 240 as the employees' bargaining agent.

A week earlier Cohen had received a progress report on union activity from a miner for whom he was conducting a compensation case:

Our campaign here is progressing fine. We have received the majority from five mines. One of the mines [i.e. Teck-Hughes] has been approached and at present we are now pending a meeting with some of the Directors. We have only one more large producer to canvass authorization from its employees. It is the Wright-Hargreaves mine. We considered it the toughest and left it to the last. The results at other mines will assist us to get the majority at this one.[3]

When Cohen learned that the union's support was growing and that the employees at these other companies were interested in pursuing a course of action similar to the one adopted at Teck-Hughes, he advised the union that it could have one board deal with all the disputes, or at least with a common chairman. The union apparently decided to make a test case of the Teck-Hughes dispute.

At a meeting of the employees, a secret ballot strike vote was taken. The results were 332 favouring strike action, 68 against, and 18 spoiled ballots.[4] The employees also authorized two employees to seek a board of conciliation. On 28 November 1939 the federal Department of Labour received an application for the establishment of a conciliation board signed by John R. Bland and Ernest Fisher, acting on behalf of the 'underground miners, millmen, mechanics, surface workers, steel sharpeners, hoistmen, cage tenders,' and others employed at the Teck-Hughes mine at Kirkland Lake. The application stated that 510 of 628 employees had signed a petition authorizing Local 240 to represent them in negotiations with the management of the mine, and to seek an increase in wages of at least fifteen cents an hour to compensate for the increase in the cost of living. After the petition had been circulated, a committee consisting of Tom Church and Nick Rozok, interviewed J.G. MacMillan, the superintendent of the Teck-Hughes mine, in order to arrange a meeting between management and a committee chosen by the employees. MacMillan told them he would inform the board of directors and advise them of the board's position. The board of directors refused to meet with the union committee.

Both J.L. Cohen and Charlie Millard (director of the Steelworkers' Organizing Committee (SWOC), and a supporter of the union's organizing objectives) were concerned that the procedures followed might still be inadequate to ensure the establishment of a conciliation board. Cohen correctly expected a company challenge. When the company received a copy of the application for a conciliation board, MacMillan protested to the Department of Labour about the manner in which the strike vote had been taken.[5] He alleged that the vote did not indicate the true views of the employees, and that Local 240 only represented a small minority of the employees. He claimed that the petition was simply a request for an increase in wages, that employees had been solicited to vote by union organizers, that the balloting was improper, and that untruths about the company had unduly influenced the voting.

In order to resolve conflicting views about the extent of Local 240's support, the department decided to conduct a secret ballot of all the mine employees. The union approved this procedure but urged that it take place without delay so that the companies would not have the opportunity to take reprisals against union supporters. The union was concerned that company threats might create a coercive atmosphere in which the true wishes of the employees were not likely to be ascertained.

On 31 January 1940 the company, without warning, discharged forty-seven employees, including the two men who signed the application for a conciliation board. Bob Carlin, a union activist and later financial secretary of the local union, was among those fired. The union was aware that the mining companies, through their spies, had knowledge of the identity of the union members, and complained to the deputy minister of labour that the discharges were motivated by these employees' union activity. The deputy minister responded by advising the company of 'the seriousness of the situation caused by the dismissal of the men at the time,' and stating that such conduct might necessitate the establishment of a conciliation board without taking a preliminary ballot.[6] On 2 February 1940 the department's chief conciliation officer (Campbell) and his assistant went to Kirkland Lake, and five days later the vote was held.

Employees were asked to vote yes or no to the following statement: 'Failing an adjustment of the dispute now existing or a reference thereof by the Minister of Labour to a Board of Investigation and Conciliation, I hereby authorize the calling of a strike.'[7] Of the 657 persons eligible to vote, 632 cast ballots: 383 (61 per cent) in favour of strike action, and 247 against, with two ballots spoiled. Clearly there was a high level of awareness about the dispute and the local union enjoyed substantial support despite the employers' recent actions. The vote guaranteed that a board would be appointed. The department decided to establish a board of conciliation composed of the Honourable Mr Justice W.M. Martin as chairman, Mr G.C. Bateman for the employers,[8] and Mr J.L. Cohen for the employees.

The board met in Kirkland Lake on 6 and 7 March 1940 and dealt first with the discharge cases. The union again complained that the firings were intended to eliminate union supporters, intimidate other employees, and induce them not to join the union. MacMillan reiterated the company position: that tonnage had been reduced and the forty-seven 'least efficient men' had been laid off. He pointed out that while thirty-four of those dismissed were union men, thirteen were not. The union believed the presence of non-union men among those dismissed was to camouflage the company's intention to discriminate, but in any event, MacMillan's remarks indicated the employer's detailed knowledge of union activity. The board's findings on this issue concluded that 'considering the circumstances, the method and time of dismissal and the number of union members dismissed, including Bland and Fisher, it is not surprising that some of the employees were of the opinion that at least some of those discharged have been dismissed because they were members of and active in the union.'[9]

The board did not question the management prerogative to reduce tonnage and lay off men, but indicated that the union's concerns about discrimination and the failure to respect seniority were valid. The board noted that 'seniority is of course an important consideration and should apply whenever possible.' It also felt that the timing of the lay-offs of so many union men just prior to the strike vote was 'ill advised.' It 'would not tend to bring about a settlement of the matters in dispute; on the contrary it would only tend to increase the tension which already existed.'[10] This moderate tone in the face of the clear evidence of anti-union motivation can perhaps be explained by the board's inability to provide any remedy other than an unenforceable recommendation.

The board recommended that six of the discharged men be reinstated at once (including Fisher and Bland), and four others selected by MacMillan be rehired. This was done. Bob Carlin was not rehired but subsequently was hired by Mine Mill as a paid organizer. The company agreed to re-employ nine men immediately and others as needed. The board recommended that future lay-offs and rehiring be done according to seniority. Nevertheless, according to the union, this incident became a positive incentive for trade union organization – perhaps because the company's high-handed action indicated more clearly than anything else the true nature of its relationship with its employees.

The next issue the board dealt with was the question of union recognition. Management objected to recognizing Local 240 of the IUMMSW (affiliated to the CIO in the United States and the CCL in Canada) on the grounds that 1) the local was part of the CIO, and 'our observations of the methods and tactics of the IUMMSW [Mine Mill] in Ontario, and other provinces leads us to the belief that it is an organization in which little confidence can be placed'; 2) the local was dominated by the Communist Party; 3) the local's organizing campaign deliberately included untrue statements about the company, circulated to create dissatisfaction among the employees. The companies took essentially the same position in 1941.

These submissions had little impact on the conciliation board. With regard to the CIO affiliation, and the management charge that the CIO was an irresponsible organization, the board merely noted that 'it is a fact that affiliates of the CIO have won very large support in the U.S. and have received very wide recognition by industries. Some of the largest employers of labour have made agreements with these unions.' The board also rejected the company's charges of union intimidation and misconduct during the organizing campaign 'as no such methods have appeared in the dispute at Teck-Hughes Gold Mines Limited, the Board would not be justified in finding that the methods and tactics of the IUMMSW leads to the belief that it is an organization in which confidence cannot be placed.'

The board was persuaded that the union's estimates of company profits and the number of accidents had been 'greatly exaggerated' and that such misinformation prejudiced the possibility of a harmonious relationship. The board found little merit in the company's charge of communist domination. Tom Church, the international union board member and former local union

president, had the support of the communists in his unsuccessful candidacies
to various posts and elections to Teck Township Council. He had 'unfortunate' past associations and had made some careless remarks. But Church and
Bland (the president of the local) denied being communists. The board concluded that 'so far as the personnel of the local union is concerned, it does
not appear that there is a sufficient reason for concluding that it is not a
proper body to be dealt with by the mine management.'[11] In fact, of course,
there was a communist element in the local; it did not affect the appropriate-
ness of the local union as a bargaining agent representing the miners, but it
did prove to be a source of disunity that had to be resolved before the local
could become an effective fighting organization.

On the question of union recognition, the board quoted former labour
minister Norman Rogers' view that unions would continue to grow whether
union recognition was legislated or voluntary. Noting recent provincial
legislation,[12] the board concluded:

In view of the general recognition which has been given both by law and practice to
the right of workers to organize and to the right of collective bargaining, it does not
seem reasonable for any industry to refuse to recognize these rights unless there is
some substantial justifying reason. In respect to this dispute no sufficient justification
has been shown to the Board for the refusal of the company to recognize local union
number 240 of the IUMMSW.[13]

The board's report was unanimous on all issues *except* union recognition.
Mr Bateman, the management appointee and a prominent person in the
mining industry, repeated the company charges in a minority report. He
believed that the local union was closely associated with the Communist
Party and he condemned the union's organizing tactics. He noted that 'an
almost unbridgeable gap exists between the union and management,' and
concluded that 'whatever the wisdom may be of the ultimate recognition of
some form of organization, my own opinion is that no good purpose can be
served by attempting to force such recognition at this time.' In his view,
forced recognition would only aggravate the situation. The United States
experience did not support this view, but the question of union recognition
was left to the future.

In May and June 1940 the board considered the union arguments for a
fifteen-cent-an-hour wage increase. The union based its case upon the high
cost of living in Northern Ontario, the dangers and hardships of the mining
industry, the difficulties of saving because of the short duration of mining

jobs, and the company's ability to pay. The company alleged that the average earnings of the men working in the Teck-Hughes mine were higher than any other Ontario industry, and that the standard of living of the miners had risen over the years. The company pointed out that in the depression years the average weekly wage had *risen* at a time when the cost of living had dropped substantially, and contended that the increased price of gold was not absorbed solely into profits, but was reinvested to increase production of lower grade ore.[14] These arguments were virtually identical to those raised in the later Kirkland Lake dispute, although by 1941 wartime inflation was recognized as an additional problem.

Ultimately the board agreed that the wage question was fundamental, but found that it involved not simply wage levels at the Teck-Hughes mine, but also wages within the entire mining industry. In order to determine whether the union demand was fair and reasonable, it was necessary to investigate the wage levels and differentials within the industry generally and to compile extensive data on both the labour and product markets. As the board could not examine the wage question in all of its aspects, it determined that it could not reach a firm conclusion on the situation at Teck-Hughes. The board estimated that the cost of living had risen 7 per cent since the outbreak of war, and acknowledged that there was no comprehensive welfare scheme for employees. It decided, however, that its term of reference did not permit an examination of these 'fringe benefits' and that, in view of the paucity of information, the entire wage question should be deferred for at least six months. The general question of fringe benefits was referred back to the parties.

The wages issue was an important one for the miners, and it would not disappear. However, there was no further inquiry into the wages of the mining industry until the 1941 Kirkland Lake strike. The only adjustments made were to the cost-of-living bonus in order to conform to the government's wage order (PC 7440). This did not affect the basic rates of the industry and was, therefore, not satisfactory to the union. The union position was expressed in its brief to the Teck-Hughes board.

I would like to be able to submit to the Board a statement of average bonuses which a worker could expect; unfortunately no such estimate is available from the employees. The reply I have repeatedly received is, 'Only the company could tell you that'; mistrust and some resentment is directed toward the whole system by a majority of the employees. They believe that it is used ... capriciously ... in the interests of favouritism and intracompany politics. It is one of those matters which we hope a

reasonable system of collective bargaining will see amiably adjusted. By any meas-
urement, we contend it is far from adequate in its professed purpose of compensating
for occupational dangers.[15]

The bonuses were calculated and distributed in accordance with company
policies that the miners did not understand and suspected were inequitable – particularly when the company refused to discuss them with the
employees' bargaining representative. To raise the bonus without additional
information about its administration did not meet the miners' wage demands
and did not dispel their suspicions, but behind these specific complaints was
the fundamental conflict over the right to collective bargaining. The miners
believed that all of these matters should be resolved by negotiation with their
employers, while the mining companies adamantly refused to participate in
this process even to the limited degree of meeting with the union to discuss
the issues.

At the conclusion of its report, the Teck-Hughes conciliation board congratulated the parties on their conduct – a gesture which could not be
repeated in the more polarized situation of 1941–42. The report was submitted to the deputy minister of labour on 7 June 1940 – five months after the
original government-supervised strike vote. The miners accepted the report,
but forty-five days later the president of Teck-Hughes mine informed the
Department of Labour that he did not accept it.

Nothing happened. There had been two strike votes; the union had
demonstrated its support; there had been substantial delay and a government-imposed conciliation board. And still there was no recognition, no collective bargaining, and not even an objective assessment of the union and
company positions that might have prompted one side or the other to compromise. In the circumstances it is not surprising that the union distrusted
the government and regarded its labour policy as a device to frustrate union
aspirations. The government had no intention of enforcing the board's
recommendations. As *Canadian Forum* magazine portrayed it at the time,
'the company contemptuously ignored the findings of the board. The government did nothing.'[16]

The significance of this early case at Teck-Hughes was that it 'convinced
the miners that it was useless to deal with mining companies individually.'
The cumbersome compulsory conciliation process had been useless. It
neither resolved the dispute nor provided a basis for negotiations. Resolution was impossible so long as the employer refused to recognize the union.
Conciliation merely delayed the inevitable conflict and gave the employer

months to prepare so that by the time the employees were in a legal position to strike, the bargaining power could be seriously eroded.

The union believed that it had failed at Teck-Hughes because its organizational basis was weak and it had put too much faith in the conciliation process. The justice of its position was quite irrelevant when the company refused to accept it and the government had no intention of enforcing acceptance. The means for achieving success lay not in developing reasonable arguments, but in strengthening the ability of the union to apply economic pressure. The international union and the local organizers in Kirkland Lake believed there were great prospects for organizing all of the mines.[17] The union, therefore, concentrated its organizational efforts in Kirkland Lake so that it could consolidate its position in *all* the mines and seek a master agreement. The government could not be indifferent to the possible interruption of the entire output of the Kirkland Lake gold mines, nor could the individual mines play one group of employees against another or fulfil their production commitments from the stockpiles of those mines that continued to operate. A master agreement would deal with all the outstanding issues including wages, working conditions, and union recognition; and it would have the effect of standardizing inconsistent wage standards within the mines. With the imposition of wage control there was little possibility of substantial wage increases, but the concern that there be 'equal pay for the same work' remained. Security on the job and the protection of the local union became *the* dominant issues.[18]

Before the organizing campaign could continue effectively, the union had to resolve the problems of factionalism and internal disunity. The first active workers in the local had been communists, but as the miners began to respond to the union's organizing drive others joined who were unsympathetic to the political views of these older members. As a result, ideological debate created dissension among the members and diverted attention from the practical tasks facing the union. As Bob Carlin described it, 'the fight was between the two "isms" – Communism and the CCF Party – a battle for supremacy.' Carlin was one of the leaders of the CCF faction and Tom Church (the international board member) was 'the key guy for the Communist Party.'[19] Both groups brought their leaflets to union meetings and raised political issues on the floor. The effect of such activities was to confuse the 'old timers' and to alienate the young people. Prospective members would attend a meeting to discuss collective bargaining and other union 'pork chop issues' and find themselves involved in a debate as to which political philosophy was the best: communism or socialism. While some people may have been attracted because of their ideological convictions (and the issue of government labour policy was undoubtedly 'political'), most miners were initially concerned about building a strong union that could force employer recognition. The political conflicts were destroying the local. Many of the younger miners interested in joining the union were unwilling to become involved in intra-union ideological battles, which seemed remote from the immediate needs and objectives of the union, yet seemed to dominate the debate at union meetings.

Since 1939 Tom Church had been the Canadian representative on the international board of the union. His communist connection had been the subject of comment by the Teck-Hughes conciliation board. He had, in fact, diverted some of the local union's funds to political activities. President of the IUMMSW, Reid Robinson (whose views were not, as yet, similar to Church's), was anxious to see the union grow, and was concerned that too much emphasis on politics would undermine the organizing drive.[20] He sent his trouble-shooter, Alex Cashin, to investigate. Cashin subsequently advised Robinson to put the Canadian district under administration and remove Church from his executive position. This was done in 1939. Consequently, the Canadian District no. 8 had no member on the international union board until Bob Carlin was elected, unopposed, to that position in 1942.[21]

As a result of these internal problems the Kirkland Lake local was barely functioning,[22] and all the disputants eventually came to believe that someone uninvolved could be of greater assistance in the organizing drive.[23] Reid Robinson was anxious to maintain the Canadian organization and was instrumental in providing this support. 'There was demand for staff people and organizational work in many parts of the United States at that time, but President Robinson felt that this area [Kirkland Lake] could not go neglected.'[24]

In early 1940 Tom McGuire was sent to Kirkland Lake as general organizer and union administrator for the Northern Ontario area. At the same time Bob Carlin was appointed organizer for the local. He had been an active trade unionist since he joined the Western Federation of Miners (WFM) in 1916, a charter member of Local 240, an executive member of the local since 1937, and one of the persons who had been fired and not rehired at Teck-Hughes in 1939. McGuire was sent in to discipline the local leadership: 'Tom came over, and he had no interest in the Communist Party, to begin with ... They called him "Red McGuire." He was a "red baiter" in the United States, so they said. I couldn't find any logical reason for calling him that. I suppose he cracked down on them some time and they said, "well, there's a red baiter," but he cracked down on the CCFers too.'[25] McGuire was an effective organizer, committed to the union. He was liked by the leaders of both factions, who agreed that his appointment would be beneficial to the local.

McGuire had originally come from Local 117 in Anaconda, Montana. In the early 1930s, he had served on negotiating committees with Reid Robinson, who at that time was president of Butte Local no. 1. After Robinson was elected president of IUMMSW in 1936, he brought McGuire onto the union's staff. Between 1937 and his assignment to Kirkland Lake in 1940, McGuire had organized in the Coeur d'Alene district of Northern Idaho, in Arizona, and in New Mexico. When McGuire arrived in Kirkland Lake he insisted that the political rivalries were 'secondary to the need of organization of the workers with the object of securing some security and benefits for the people in the mines.'[26] Carlin recalls McGuire telling the local trade unionists that 'I didn't come over here to build your goddamn political party ... if you think that you can build it out of the ashes of the union. You won't build it out of the strength of the union the way you're going, because you're not bringing any members in, but you're keeping a lot out. No young fellows want to come into this.'[27]

The political debates ceased to dominate meetings after McGuire's arrival, and the officers of the local changed. Tommy Church, having been removed from the union executive, was elected to Teck Township Council. Carlin remained a local union officer, but agreed with McGuire that there should be no more partisan politics until the local was established.[28] The local's president, William Simpson, had been allied with the communist faction but not so closely as was Church, while Larry Sefton and Joe Rankin (the recording secretary and vice-president of the local, respectively) were both young miners who were protégés of Tom McGuire and who became active in the union after his arrival. They, too, were primarily interested in building a strong local union. Later they would become political and support the CCF, but they were trade unionists first and saw political activity as an outgrowth of strong workers' organizations, not a substitute for them. The changes in personnel and in organizing tactics were intended primarily to strengthen the local, encourage new membership, and in the process counter management charges of 'communism' and 'CIO irresponsibility' that had been so frequently voiced during the 1939 Teck-Hughes dispute. For the time being the Communist Party's association with the local was considerably weakened. Nevertheless, the mining companies remained adamantly opposed to any union.

In the fall of 1940, having established greater internal stability, the union began a new organizing campaign. Organizing committees were established in practically every mine in order to solicit members and ultimately to enable the union to approach all of the managements at the same time.[29] From the beginning these small committees functioned efficiently – especially in the larger mines.[30] McGuire found that the presence of old timers who had been members of the WFM and other unions was very helpful in organizing.[31]

On Sunday, 20 October 1940, Kirkland Lake's Local 240 held a meeting in the Strand Theatre 'to inaugurate its organizing campaign among the mine and mill workers of [the] Kirkland Lake and Porcupine districts.'[32] A.R. Mosher, president of the Canadian Congress of Labour, was the main speaker at this 'kick-off' meeting. He urged the workers to follow the example of other industrial workers and organize to improve their economic conditions. This was to be the first of many such meetings featuring promi- nent trade union and political speakers, including Pat Conroy (director of organization, CCL), C.H. Millard (director of SWOC), and Clarie Gillis (a former coal miner and CCF member of Parliament for Cape Breton South). In November 1940 Conroy advised general meetings in Kirkland Lake and Timmins, and mine committee meetings in Kirkland Lake, on the practical techniques of union organization.

There were still some divisions in the local that created difficulties in organizing. McGuire referred to the Church–Rozok group as irresponsible and blamed them for impeding the organizing drive. Such divisions were unfortunate because at this time the mine owners mounted a counter cam- paign against the union's drive for members. During this period, the same split was occurring within the national labour movement. In May 1941 the leading communist trade unionist, C.S. Jackson (who was head of the UEW), was briefly suspended from the executive council of the CCL for allegedly publishing his own version of minutes of the council meetings. Jackson had spoken in Kirkland Lake after Mosher and Conroy had been there, and had been publicly critical of both leaders and of the CCL role in the organizing drive, which he considered inadequate. The CCL council believed such divi- sive actions were undermining the workers' confidence in the new congress and jeopardizing the northern organizing drive. To these trade unionists, trying to build new unions and a new central labour body, organization came *first* and on that issue there had to be unity if there was to be success.[33] This view was similar to that adopted by Mine Mill in respect to the political divisions that were weakening the Kirkland Lake local.

Besides the assistance of outside speakers, Local 240 received direct organizational and financial aid from the international union. McGuire had been sent from Denver, Colorado, as the general organizer and administra- tor for the Northern Ontario area, even though other locals in the union were also seeking organizational assistance. Throughout the organizing period and the strike, president Robinson made frequent trips to the area. The local was subsidized by the CIO, which paid McGuire's salary, and by Mine Mill, which paid the two part-time organizers (one of whom was Bob Carlin).

In October 1940, District no. 18 of the United Mine Workers of America (UMWA) assigned a full-time organizer, Alec Susner, to assist in organizational work. Susner worked among the foreign-speaking workers, 'reaching many whom it had been impossible to organize previously.'[34] Another UMWA organizer, John Stukoluk, spent two weeks in the area making a survey of membership and speaking at several meetings. If the UMWA was hoping to organize these miners themselves it came to nothing, and co-operation with the UMWA continued throughout the struggle.

C.H. Millard (secretary of the Canadian CIO) assisted the early organizing efforts by having stewards' books and union application cards printed at his organization's expense. These were the essential tools for recruiting members. In September 1940 financial assistance was received from the CCL which, at this crucial time, sent funds for 'preliminary organizational work in the Northern Ontario gold field.' By November 1940, McGuire reported to the congress that he had edited three issues of a 'bulletin,' two more were planned, two mass meetings had been held, and another was planned for the near future. The *Canadian Unionist*, newspaper of the CCL, reported that McGuire was making headway and had 'gained the respect of both the workers and the employers of the mining country,'[35] a very optimistic view of the intransigent attitude of the mine managers.

The IUMMSW continued to work in the Kirkland Lake area until March 1941 when McGuire gave the CCL executive committee an extensive report and requested financial aid to sustain a bigger campaign. He told the CCL that the international union would spend $1000 a month to assist the campaign,[36] as it apparently did.[37] He proposed a total budget of $2500 a month, the sum he thought would be necessary to successfully organize in Kirkland Lake and Timmins. McGuire reported that the organizing could be completed by July or August and requested that the CCL make up the difference between the international's contribution and the amount needed to sustain the campaign. In his view there was great dissatisfaction in the area and the time was ripe for unionization. There were approximately 5417 employees in the Kirkland Lake mines, of whom 800 were union members; in Timmins only 150 of the potential 9000 members were organized.

The request for large sums of money and the knowledge that the initial success was limited disappointed the CCL executive. After McGuire had made his presentation and withdrew from the meeting, the CCL executive discussed the matter at length and concluded that the IUMMSW was unequal to the task of organizing the mines of Northern Ontario 'because it is unable to provide either the men or the money required.'[38] In accordance with congress policy,[39] the CCL told McGuire that it was unable to finance organizing campaigns for affiliated member unions and advised that 'unless your International can undertake to finance the campaign some other arrangement should be made in this connection.'[40] At A.R. Mosher's and Robert Livett's suggestion, and apparently without informing president Reid Robinson, the congress executive invited the UMWA to undertake the task of organizing the hard-rock miners in Northern Ontario and Quebec. UMWA President John L. Lewis was advised that the area was 'ripe for organization' and that only the UMWA could effectively organize the 15 000 miners in Kirkland Lake and Timmins. It was suggested that Lewis make arrangements to obtain jurisdiction over this area from the IUMMSW.

Lewis refused. In 1935 and 1936 Mine Mill had sought the amalgamation of the IUMMSW and the UMWA, but at that time John L. Lewis had not been interested and the idea was eventually dropped when the CIO was formed in November of 1935 and Mine Mill found itself part of a thriving industrial union movement.[41] Consequently Lewis informed the CCL that the UMWA 'would not in any way want to intrude upon the jurisdiction of the IUMMSW' and suggested that the CCL and IUMMSW work out an agreeable organizing plan.[42] Reid Robinson, when informed by Lewis of the CCL's overture, took a diplomatic position. He thanked the congress for its interest in the northern area and acknowledged that a transfer of jurisdiction might accomplish the necessary organizing goals, but noted that the UMWA did not desire the jurisdiction. Mine Mill had recently decided to increase its spending in the area and UMWA District no. 18 had pledged its continued support. Robinson concluded: 'I am sure that through our expansion and through the cooperation of District 18, we will be able to accomplish the organizational success ...'[43]

In April 1941, Carlin wrote an article in the *Canadian Unionist* that repeated the grievances that had surfaced in the Teck-Hughes dispute. Carlin described for his fellow trade unionists the conditions that were inexorably drawing people into the union. It was neither communists nor outside agitators from the CIO, nor foreign revolutionaries, who were inducing workers to support the union; it was concern for the circumstances under which they worked. Carlin described the state of his union's organizing drive in April 1941:

All that may be said regarding wages, working and social conditions of the Northern Ontario gold miners is thoroughly understood by the miner himself. The long-stand-

ing bad working conditions, low wages, and hazardness of the occupation add impe-
tus to the organizational campaign being carried on by the IUMMSW, which has offices
and an ever-growing membership in the two largest gold-mining towns in the North.
The miners are organizing and rallying to their union in greater numbers than at any
time in its history.

Changes would only come about 'by the organizing efforts of the miners
themselves. That is why a great influx into the union is taking place.'[44]

Carlin hoped that the organizing of the metal miners of Northern Ontario
would be completed by the summer of 1941, but he was overly optimistic. In
May there were over 1000 members in the local. In June 1941, the *Canadian
Unionist* incorrectly reported that the membership of IUMMSW Local 240
encompassed 75 per cent of the miners in the area – a level of organization
that was not, in fact, achieved until some months later. The organizational
activity in Kirkland Lake spread into nearby areas as well. The union in
Larder Lake was gaining strength daily, and a union office had been opened
to further this activity. The provincial government was kept closely informed
about the extent of organizing in Kirkland Lake and vicinity through reports
by police commissioner Stringer and from secret agents stationed in Kirkland Lake.[45] Union meetings were held and on 10 June the union requested
negotiations.[46]

In early July, Bob Carlin and Larry Sefton – the financial and recording
secretaries of Local 240 – took an accounting of their 'members in good
standing' and decided that it was time to initiate bargaining with the employers. A meeting was organized to inform the membership of the local's
strength and discuss the action that their officials proposed. The first step was
to contact the mine operators.[47]

The employers were contacted by telephone on 7 July and by mail on 9
July, and informed of the union's desire to negotiate a master agreement
that would establish wages and working conditions in all twelve Kirkland
Lake mines. This request was refused, as was expected. Thereafter, there
were repeated but unsuccessful attempts to enter into negotiations with the
mine operators. In contrast to its position in the Teck-Hughes dispute, the
union was now confident that it could strike if it became necessary to do so.

4

The Industrial Disputes Inquiry Commission and the one-day holiday (July–August 1941)

As the organizing campaign reached a climax and the strength of Local 240 continued to grow, the government began to regard the gold mines in Kirkland Lake as a potential 'trouble spot.' On 21 June 1941, 'when the rumour of a possible strike occurred,'[1] the government sent its chief conciliation officer, Mr M.S. Campbell, to Kirkland Lake. He tried to convince the mining companies to meet with the union and suggested that the union be allowed to cross-check its membership list with the mines' payroll lists in order to accurately assess the extent of its support. The union suggested that the federal Department of Labour persuade the companies to accept the results of a vote by secret ballot. All of these proposals were rejected. When it seemed unlikely that negotiations would take place, the union decided to apply for a conciliation board.

In order to make such an application, the Industrial Disputes Investigaton (IDI) Act required that the union first present the companies with a list of demands. On 9 July 1941 Bob Carlin sent a registered letter to all the mine operators, outlining the union's Ten-Point Program and asking for a meeting to conclude a *master* agreement. He requested a reply within ten days. The union demands included union recognition, adjustments to wages, hours, holiday pay, and existing medical plans, a system of seniority, a grievance procedure to redress alleged violations of the agreement, provisions for 'contract re-openers' in the event of changes in the economic situation, a master agreement with supplementary agreements for each mine, and a job-return guarantee for those in military service. The mine operators ultimately rejected all of these proposals, after some preliminary skirmishing.

The Department of Labour suggested (and the companies wished) that the union not only outline points of discussion but also set out in detail the substantive position of the union on each issue.[2] The union was urged to submit a complete draft of its proposals, including detailed language so that the department might be able to secure agreement on at least some of the clauses. The implicaton was that if the union compromised a little, there need be no dispute and the parties need never meet. Of course, no collective bargaining would take place either. Carlin replied that it was futile for the local to clarify its position when the operators refused to meet and bargain or to recognize the union as the legitimate bargaining agent for the employees. In any event, the operators had already advised the press that their position on this issue was the same as it had been in the Teck-Hughes case. Carlin said that 'the operators have repeatedly refused to meet with the union committee in joint conference and also individually, so it is very easy to see that this move of theirs is a subterfuge to stall.'[3] Local 240 was prepared to accept the intervention of a department official who could bring the parties together to discuss proposals or secure counter proposals from the operators – so long as such action did not unduly delay the conciliation board.

These interchanges created further bad feeling between the parties. Charles Harris, spokesman for the mine operators, charged that there was an 'intentional slowdown' in the mines and 'definite evidence of intimidation.' Union administrator Tom McGuire denied the charge and stated that the men were actually working harder in order to avoid giving the companies any evidence of a slowdown.[4] Finally, on 18 July 1941, the union formally applied to the federal Ministry of Labour for the appointment of a board of conciliation to investigate and make recommendations about hours, wages, conditions of work, and union recognition. The local expected that the board would be promptly established, for McGuire wrote to President Reid Robertson on 24 July 1941 to inform him that it would be necessary for Ben Riskin of the international office to prepare a brief for the board hearings. He advised that Riskin should come to Kirkland Lake as soon as possible for this purpose.[5]

It was becoming a common practice when an employer refused to deal with a union for the latter to request a conciliation board. The establishment of a board put pressure on the employer and focused public attention on the issues in dispute. In the case of a recognition dispute, the board's hearings might be the only time the two parties actually met. That experience in itself implied a kind of *de facto* recognition; however, government intervention of this kind was not *necessarily* advantageous to trade unions. In the National Steel Car dispute in Hamilton, it had led to the appointment of a controller whom the union assumed, incorrectly as it turned out, would recognize the union and negotiate a collective agreement. In the Peck Rolling Mills case, a conciliation board had rejected union wage demands as being contrary to the recently enacted government wage policy. The delays involved in the conciliation procedure often prejudiced the union's chances of winning recognition or achieving a collective agreement.[6]

In early 1941, the president of the Trades and Labour Congress (TLC) had complained about these cumbersome conciliation procedures. In its annual presentation to the government the CCL had advocated introduction of the Wagner Act machinery to Canada. Convinced by discussions in the National Labour Supply Council (NLSC)[7] that their proposals would not succeed, the CCL urged that the government accept a 'short-cut device' designed to achieve speedy settlements of certain limited types of industrial disputes without the need for establishing conciliation boards. It also proposed that government commissioners hold powers to investigate charges of discrimination on account of union membership. When Aaron Mosher, CCL president, resigned from the NLSC to protest labour's dissatisfaction with the way in which the government's wage policy was being implemented, the government took action.[8] It established a new administrative agency, the Industrial Disputes Inquiry Commission (IDIC), to operate under the direct authority of the minister and to perform investigative functions more quickly than the *ad hoc* tripartite conciliation boards. Ironically, to the chagrin of organized labour, the operation of the IDIC did not speed up the conciliation procedure, but in fact added to the delays involved in that process.

PC 4020, creating the IDIC, was proclaimed on 6 June 1941, immediately prior to the Kirkland Lake miners' request for a conciliation board. In July 1941 the powers of the IDIC were expanded by PC 4844: it was given authority to investigate cases of alleged discrimination in employment because of trade union activity, or union coercion to induce employees to join a union. However, there is a considerable difference of opinion as to the effectiveness of the commission. In H.D. Woods' opinion, a large number of dismissal cases were investigated by the IDIC; in many such cases, the commissioners ordered reinstatement with back pay. This possibility of direct relief to the individual employee supported by government sanctions foreshadowed the later system of legislated unfair labour practices.[9]

The Wagner Act in the United States had included an independent agency with broad powers to remedy unfair labour practices and to require parties to rectify their unlawful acts. The remedial authority of the IDIC represented a tacit recognition of the Wagner Act principles, which ultimately were embodied in the 1943 Ontario Collective Bargaining Act, and in the federal government's 1944 order, PC 1003. The IDIC appeared to be a real concession, but as a lawyer operating in the system at the time, J.L. Cohen was sceptical. Ultimate enforcement authority was vested in the minister, and was discretionary. The government refused to enforce the principles of PC

2685 even in industries that it controlled; PC 2685 declared that 'employees should be free to organize into trade unions free from any control by employers or their agents ...' and 'employees through the officers of their trade union ... should be free to negotiate with employers.' It is, therefore, understandable that in 1941 Cohen could write that 'the protection afforded by PC 4844 in cases of discrimination is quite illusory' as 'no forum is designated, no administrative process is provided, much less is any penalty provided.'[10] PC 4020 provided both the authority and the mechanism for government action, but it was not *required* to entertain complaints of employer misconduct or provide a remedy. By this time the labour movement had little reason to trust the government, or believe that the new agency was anything more than another empty gesture.

An examination of the terms of PC 4020 provides further support for Cohen's scepticism. Under the IDI Act, the parties were entitled as a *right* to a board of conciliation, which was the essential prerequisite to strike action and which, in recognition disputes, provided the only forum where union and management were *required* to face each other and state their positions. For the purposes of conciliation, the status and legitimacy of the union as the employee's bargaining agent could not be questioned. The structure of conciliation recognized the *right* of the union to appoint a nominee to the board and appear before the board. In this sense the conciliation process conferred a form of union recognition, and the trade unions were anxious to proceed to this stage without delay. PC 4020 was allegedly intended to expedite the process and purported to be a sympathetic response to labour's expressed concerns. However, this intention is by no means clear from the language of the order.[11] It would appear that the right to conciliation had been qualified and made subject to the recommendation of the IDIC, which presumably could recommend that no conciliation board be appointed. But conciliation remained a prerequisite for a lawful strike. PC 4020 seemed, therefore, to have restricted employees' right to strike.

In recognition disputes, it was not clear how the IDIC could achieve a 'mutually satisfactory adjustment,' since recognition is not an issue that is susceptible to compromise or a middle position. With the existence of PC 2685, one would have thought the IDIC could only succeed if it could either persuade the employer to recognize a trade union or persuade the workers to abandon their claims to trade union recognition. As it turned out, the IDIC adopted the latter approach.

At the outset the role that the IDIC was to play in this process was ambiguous, since it was forbidden to 'offer any opinion as to the merits or substantial justice of such features of the case as may have to be submitted to a board of investigation and conciliation.' At best it seemed that the IDIC could act as a channel of communication between the parties since no third-party recommendations were permitted; at worst (from labour's point of view), it could deny conciliation altogether. It appeared that either the IDIC would usurp the functions of a conciliation board, or it would be largely redundant. It could not be viewed as a mechanism to expedite the conciliation process. Thus, although the government purported to respond to organized labour's concerns in passing PC 4020, the actual terms of that order support a contrary inference. Labour's suspicion of the government's motives was deepened by the actions of the IDIC in Kirkland Lake and elsewhere. The IDIC was no more successful at resolving recognition disputes than the conciliation boards had been, and it came to be regarded by organized labour as a device for imposing government-sanctioned company unions.

The parties in Kirkland Lake were informed on 2 August (three days prior to the deadline set for the minister of labour to establish a conciliation board) that the IDIC would investigate the dispute and that, therefore, no conciliation board would be constituted. The union had been warned that this might happen and on 23 July had written to the government protesting the need for any further investigation. Conciliation Officer Campbell had made a thorough investigation and had assured the local that he would recommend the *immediate* appointment of a conciliation board without the necessity of a strike vote. The Department of Labour ignored the local's protest and advised that the IDIC would be in Kirkland Lake between 5 August and 7 August. Tom McGuire reflected the union's exasperation: 'We don't like this one bit, and will take a very firm stand on this investigation, particularly after Campbell ... made a thorough investigation when here and recommended a Board without delay.'[12] As far as the union was concerned, the federal government had deliberately adopted a policy of procrastination that would impede organization. There is some indication, however, that the government honestly believed that the intercession of IDIC Commissioner Humphrey Mitchell might 'settle' the dispute.

By the time of the IDIC's intervention in the Kirkland Lake dispute, it was becoming clear that this commission had adopted a policy of attempting to persuade workers to abandon their trade unions and their demands for recognition. Beginning with its first union recognition case in the Canadian General Electric plant, the IDIC appeared 'to be assiduously attempting to remove the matter of trade union recognition and collective bargaining from the arena of labour disputes by cultivating acquiescence of workers to one-plant committees, the employer's idea and model of labour relations.'[13] This approach, which labour correctly saw as undermining independent unionism, was evident in the published account of the activity of the commission in Kirkland Lake. Its proposal became known as the 'Kirkland Lake Formula' and enjoyed a certain notoriety among trade unionists. The same approach was taken by the IDIC *after* the Kirkland Lake strike, in a dispute at Canada Packers Limited. While the PC 2685 principles were the expressed public policy, employees' committees became the preference of this important government agency.

The idea of employee committees (as opposed to broadly based independent unions) had always appealed to Prime Minister Mackenzie King. His ideas about trade unions did not seem to change substantially throughout the years,[14] and his views were not merely expressed in his writings but were implemented in his Colorado Plan. King's consistent aim in labour relations was to avoid industrial strife and unrest. He feared the adversarial nature of industrial relations, as was particularly evident in the war years, and preferred employees' committees because they played an accommodative role. The failure of the government to recognize the legitimacy of employees' interests and its belief that they were no different from the interests of employers help to explain the tenacity with which the government pursued conciliation, consultation, and mediation in preference to collective bargaining in its handling of the National Steel Car dispute, its enunciation of the Kirkland Lake Formula, its delay in legislating the recognition of unions in Crown corporations, and its tardy proclamation of PC 1003 (enforcing compulsory collective bargaining) only after extreme pressure was applied. The government viewed the assertion in industry of a separate, independent, and conflicting employee interest as illegitimate and designed its policy accordingly. Humphrey Mitchell agreed[15] with Mackenzie King's conception of industrial relations. This accord was the basis for his promotion first to the head of the IDIC and later to labour minister. It also was the basis of labour's distrust of the actions of both men. The IDIC was used in recognition disputes to promote a government policy that was not neutral, but rather sought to promote industrial peace at the expense of collective bargaining.

In early August, at the request of labour minister McLarty and IDIC chairman Mitchell, commission members George Hodges of the Canadian Pacific Railway (CPR) in Montreal and Professor Gilbert Jackson of the University of Toronto ('one of the most conservative economists in Canada')[16] met in the Park Lane Hotel in Kirkland Lake with a group of employees (which included union secretary Bob Carlin and district administrator Tom McGuire), and then, separately, with the mine managers. At the time the local paper, *Northern News*, commented that the IDIC was faced with a difficult assignment, for the union viewed any device short of a board of conciliation as stalling on the part of the federal government.[17]

The meeting of the commission with the employees and union officials apparently accomplished nothing. The union briefly questioned the commission's authority, but to no avail. According to the commission's report to the Department of Labour, 'there was little discussion other than in respect to the nature and cause of [the] dispute which in the application for the establishment of a Board of Conciliation and Investigation was stated to be the refusal of the mine operators ... to meet with the representatives of the Union to negotiate a collective master agreement and recognition of the Kirkland Lake Mine and Mill Workers Union, Local number 240.'[18] Some references were made to the union's economic demands but the commission was made aware of the employees' *overriding* concern for union recognition and collective bargaining rights.

The companies also clarified their position. They had never had any written agreements with their employees before, and they were, therefore, disinclined to enter into any such agreement now, especially with a CIO organization. They denied the need for a conciliation board since their *employees*, as distinct from the union, had not approached them 'directly' to discuss wages and working conditions.

The commission explained its interpretation of the government's wage policy and PC 2685 to both parties. In its view, compliance with the wage policy might require adjustments to the bonus system but not to the basic wage rate; and the requirements of PC 2685 could be met by employer recognition of employees' committees, but not of a union. The union rejected the IDIC interpretation. Apart from its main demand for union recognition and collective bargaining, McGuire told the commission 'in no uncertain terms' that the employees would insist on an increase of fifteen cents 'right across the board' and would enforce this demand by striking if necessary. The union believed adequacy of wage rates should be judged by whether or not workers could live at a 'decent' standard of living, and that automatic pegging of wages meant that workers were locked into substandard levels. This conflict in the interpretation of the wage order was similar to the one that had occurred at Peck Rolling Mills. However, the wage issue became less important as the dispute continued.

The companies presented the same arguments that had been made at Teck-Hughes and would later be presented to the conciliation board. The company position was that the rates of pay were the highest in the history of the industry. They had last been raised in 1936 by forty cents a day. The basis of evaluating the level of wages under the government wage policy had been met. There had been upward adjustments of individual rates in various classifications. In January 1941 the companies had also initiated the payment of a cost-of-living bonus of $1.50 a week. (They had deliberately introduced this change before the government wage policy was proclaimed in PC 7440. Consequently, as the IDIC discovered, this bonus had not been calculated in accordance with the order. Had it been so it would have amounted to a wartime bonus of $1.80 a week.) Because the wage levels were fair, the company submitted that recognition of the union, and collective bargaining, were not serious issues.

The companies pointed out that they were caught in a 'cost-price' squeeze because of higher wartime costs and more production of low-grade ore. Some of the low-grade mines were working on a low margin of profit because the selling price of gold was fixed. Even the larger mines were working low-grade ore, in some cases at a loss. Such mines would cut back on production of these ores if their costs rose. Any increase in costs would result in closures. Less gold would be available for export to the United States, and fewer miners would be employed. The IDIC was receptive to the companies' presentation. The later conciliation board would never get the opportunity to make a judgment on the wage issue, but would only deal with the issue of union recognition.

As a result of its meetings, the IDIC got a specific undertaking from the mine operators that accorded with its interpretation of the various aspects of the government's labour policy and which it believed represented a substantial shift in the management position. On the basis of this undertaking, the IDIC made a proposal for the settlement of the dispute. This was quite improper, since the language of PC 4020 suggested that the IDIC was purely investigatory and provided 'that the Commission shall not offer any opinion as to the merits or substantial justice' of a case. The IDIC admitted to the government that it had told the union that the draft proposal was 'of *their* composition and that they thought it a fair proposal to settle the pending dispute.'[19] The IDIC had, in fact, urged the union to accept it, and denied that it had adopted a management proposal. Nevertheless, it did associate itself with the proposal in a way that advanced the employer's cause, as is demonstrated by an article in the *Northern Miner*: 'From the standpoint of the union, the report of the Humphrey Mitchell Commission was a distinct disappointment. The suggestion that the principle of collective bargaining be followed through the medium of mine committees rather than by recognition of the CIO union did not suit their book.'[20] While it is understandable that the IDIC would reach some conclusion as to the merits of the dispute, despite the terms of PC 4020, the unsettling aspect of the decision in Kirkland Lake, in labour's view, was that it was not consistent with the standards and principles of the government's own declaration – PC 2685, which endorsed the rights of trade union organization and collective bargaining.

The IDIC recommendations represented an alternative to the union, which was a condition not acceptable to the labour movement.

The IDIC proposal had a number of elements:

1 The companies agreed to negotiate and enter into a one-year agreement with 'committees of employees of *each* company' that would 'govern the rates of pay and working conditions' of the miners. Thereafter such contracts would be revised with thirty days' notice given by either party. The committees were to be elected by secret ballot by the employees before 31 August 1941. The elections were to be arranged by the companies' departments and supervised by a teller for each side. To ensure uniform elections among the companies, and if a company defaulted in its task, the parties could ask the IDIC to appoint a labour department officer to supervise this procedure.
2 The mining companies agreed to raise their cost-of-living bonus of $1.50 to $2.40 from 1 July 1941 in accordance with the government's wage policy. The increase was to be based on the rise in the cost of living index from August 1939 to June 1941 – an increase of 9.6 per cent.[21]
3 It was understood that if this agreement were accepted by the union, the application for the conciliation board by the union would be withdrawn.

On 7 August, at a meeting with the IDIC, this draft proposal was submitted to the union negotiating committee. Tom McGuire speaking on behalf of the employees told the commission that the companies' undertaking, recommended by the commission, 'would be entirely unacceptable to the employees.' Both the parties to the dispute knew that employees' committees were a device to undermine the union and frustrate collective bargaining. The proposal not only separated the miners from their union, but also from each other, since it rejected the concept of a master agreement and fragmented the bargaining group. The union refused to consider the proposal and on 10 August 1941 recommended rejection of the IDIC report:

Your committee wishes to point out ... the fact that it [the report] suggests Departmental Committees or Employee Committees to negotiate with the Managements, endeavouring to divide the workers of Kirkland Lake into many parts and under such a set-up that they could control. In plain words, the draft suggests that the workers of Kirkland Lake accept a Company union in preference to a bona fide legitimate Miner's Union.[22]

The membership voted unanimously to reject the Kirkland Lake Formula of the IDIC as being an insult and 'contrary to the principles of the labour movement.'[23] The union indicated that it had no intention of retracting its application for a conciliation board.

The IDIC apparently felt that the companies' undertaking was a sufficient change in the circumstances to warrant recommending *against* the appointment of a conciliation board. However, as Mr Campbell and the Department of Labour had previously promised a conciliation board, the IDIC understood that 'some criticism might arise were a Board of Conciliation and Investigation not established.'[24] But notwithstanding the union's rejection of its report, the IDIC suggested to the Department of Labour *that the employees' committees be established and that negotiations between such committees and the companies go ahead.* Instead of seeking a 'mutually satisfactory adjustment' of the dispute, the IDIC was actually advising the government to *impose* a settlement that ignored one party. The IDIC proposed that if all issues in negotiations between the mine employers and the committees were not then resolved, the remaining questions could be referred to a board of conciliation. Such an agenda represented even further delay and an implicit rejection of the principles of PC 2685. The commission further suggested that the minister of labour 'require' the union officers to go to Ottawa so that he could urge the union to accept the IDIC recommendations and abandon its application for conciliation. The IDIC's view that conciliation of the dispute was unnecessary is unusual, since the conciliation process was a forum for voluntary resolution of disputes and the commission did not have any means of enforcing its recommendations. Of course, conciliation was also a step the union was required to take before it could apply direct economic pressure on the employers. The IDIC apparently did not take the union's opposition seriously, or it believed that it could be pressured, or perhaps even ignored. Through the establishment of an alternative to independent unionism, the IDIC was attempting to promote the government policy of industrial peace at the expense of the union's existence.

However successful this approach may have been in avoiding industrial conflict, it was neither 'neutral' third-party intervention, nor consistent with the principles of PC 2685. The IDIC did not approve of the current bargaining agent, so in the interests of industrial peace the employees should choose another. The IDIC did not approve of a common front or a master agreement, so the employees should bargain separately. The IDIC did not think conciliation was necessary, so conciliation should be delayed. The commission asserted that its solution was within the spirit and intent of clause 7 of order-in-council P.C. 2685. If the Kirkland Lake Mine and Mill Workers Union 240 have the membership they claim, in each of the mines, the union should be able to elect members to represent the majority. If after the election of committees of employees and following negotiations between such committees and the companies, there remain any questions in dispute, such specific questions if desired *and justified*, could then be matters for reference to a Board of Conciliation and Investigation.[25]

However, this approach would remove the problem of collective bargaining from the jurisdiction of the conciliation boards. As one newspaper put it, 'none of the demands contained in the original schedule have been treated in the proposals submitted by the Industrial Disputes Commission, which makes no mention of the union in any way.'[26] To J.L. Cohen, the union's lawyer, the IDIC report was

a complete renunciation of any claim for trade union recognition and trade union representation and that it proposed ... the adoption of the discredited one-plant committee plan ... It represents a formula which violates every accepted principle of collective bargaining, to say nothing of the specific terms of P.C. 2685, the order-in-council to which, by the terms of P.C. 7440 'all agreements negotiated during the war period shall conform.'[27]

It was also a solution that ignored the employees' own experiences. The Teck-Hughes board had noted the interrelationship of wages among the various mines, and the union in that dispute had faced the combined opposition of the employers. During the Kirkland Lake strike, the operators established committees in Kirkland Lake and Toronto to co-ordinate their opposition to the union. This co-operation among employers, and consequent increase in their bargaining power, was one of the reasons that the union was determined to maintain a common front and seek a master agreement. Yet the IDIC favoured separate committees.

What was particularly challenging both to Local 240 and to the labour movement as a whole, and what eventually made the strike a national issue, was the commission's interpretation of PC 2685. The labour movement believed that the IDIC report represented the re-emergence of Mackenzie King's notorious Rockefeller Plan, which had delayed unionism in Colorado for eighteen years.[28] The committees proposed in the Kirkland Lake Formula were similar to the employer-dominated committees that had been established throughout the United States and Canada as a device to undermine rather than promote collective bargaining. As J.L. Cohen pointed out, perfectly expressing the view of the labour movement:

The essential feature of collective bargaining, without which it cannot be said that collective bargaining exists, is the independence of the bargaining medium operating on behalf of the workers so that it meets on *equal* terms with the employers. Anything which destroys that independence violates the first essential of collective bargaining. Any form of employee representation or employee recognition which destroys the independence of the bargaining medium, and which is acquiesced in by the workers, not on their own volition, but only because of the influence, or control, or dictate of the employer, is a *collective bargaining form which emanates from the company and not from the worker*. That form of arrangement, whatever the variation, is therefore a form of company unionism.[29]

Fragmented, isolated, and weak, the employee committee could easily come to be dominated by management, particularly when employers were still able, with relative impunity, to discharge or pressure those employees who opposed them. Without the assistance of qualified staff and the support of fellow employees in other plants, the imbalance of bargaining power would make real collective bargaining virtually impossible. Unlike that of the skilled workers of the AFL and the TLC, the bargaining strength of industrial workers depended upon their establishing large industrial unions. The one-plant committee was a mechanism 'designed to deprive workers of the aid or support of outside representation.'

The committee solution also had historical precedents in the mining industry,[30] and given the mine operators' attitude to industrial relations, it is not surprising that they were prepared to agree to employees' committees as an acceptable forum for collective bargaining. The mine operators were under no illusions about how such systems would *really* work. Employees would not be free to negotiate on their own terms, having to function in a situation in which management was in control.

After the release of the IDIC's report, the mines did set up employee committees. Elections for this purpose were held on 31 August 1941, but the operators were able to enlist the support of only a minority of the workers. When one considers the way in which the committees operated, it is easy to understand the union's scepticism about the independence of the proposed employees' committees.

The constitution of the 'workmen's council' at Lake Shore Mines Limited, for example, set out a novel method of adjusting individual grievances. Complaints would be heard first at the employee/foreman level, then at the superintendent level (where the workers could have the support of an elected representative), then at the workmen's council level, and if there was still no solution, 'the matter [could] be presented to the Manager for definite settlement.'[31] Presumably, if the employee was dissatisfied with management's disposition of his grievance, he could quit.

The manner by which the employees were to choose 'their' representatives is also instructive. In its first year, the *company would appoint* a committee that would nominate employees from each department to run for the workers' council. Employees would *then* be given the opportunity to vote on this selection of nominees for their council. The items the workers' council could discuss included working conditions, welfare, safety, but *not* wages, which remained the prerogative of the mine management.

Besides lending legitimacy to genuine independent employees' committees, of which there were few,[32] and to company unions, the IDIC report raised the possibility that in other situations it would be acceptable to by-pass a union and assist an association competing for recognition as the bargaining agency for either all *or some* of the company's employees. Even if a union gained recognition, the government's approval of employees' committees suggested a union would have difficulty becoming the *exclusive* bargaining agency for a group of workers. This divided the employees, reduced their bargaining power, and created a situation where the employer, through threats, promises, or favouritism, could undermine the independent union. Several years before, the United States had adopted the principle of an *exclusive* bargaining agency for the union with majority support, and banned employer-dominated organizations.[33] The issue was an important one for unions and one to which they were very sensitive.

After the IDIC report was made public, it was condemned universally in labour circles, but Mitchell's private correspondence indicates that he rejected the interpretation that his solution was tantamount to 'company unionism'; or at any rate, not 'the form of company unionism implied in the leaflet'[34] put out by the Fellowship for a Christian Social Order (a pro-union church group based in Ontario). He defended the IDIC, which he believed was operating under the severe pressure of work, and outlined the commission's role as he saw it, which was

to have the employer go as far as he can possibly be persuaded to go towards meeting the desires of the employees. If the employer will not agree to recognize a union ... as a bargaining agency for the employees, our endeavour is to get him to go as far as possible towards establishing a collective bargaining relationship with representative employees, appointed by the employees by whatever method they may adopt, and with complete freedom on their part to affiliate with any labour organization of their own choice, if they so desire.

In practice, the commission was assisting the employer to circumvent a union since it was not called in until after the employees had organized. To be fair, of course, the IDIC was working as an agency of a government that had consistently rejected a policy of compulsory collective bargaining.

Mitchell's definition of recognition and collective bargaining was, in his view, a flexible and pragmatic one. In various situations the commission had assisted in bringing about settlements with various forms of recognition,

as was only practical. They have varied from outright recognition of the union by the employer; recognition by the employer of the union committee as representing the employees; recognition of the employee committee and the right of the organizer to sit at conferences to assist in the negotiation of an agreement on behalf of the employees; in some instances the employer, and I think rightly so, has insisted on a vote among the employees to determine the composition of the committee to represent them in negotiations.[35]

None of these alternatives related to the Kirkland Lake case with the possible exception of the last one. In Kirkland Lake, *new* committees were to be elected, notwithstanding the presence of duly elected union negotiating committees.

Mitchell argued that it was a major concession for individualistic managements to agree to negotiate individual written agreements with employee committees that might be composed in part of union members. The commission tried to convince McGuire that this arrangement might be the first step toward collective bargaining with the operators. If the union could get its people elected to the employees' committees and show 'responsibility' in negotiations, Mitchell apparently believed that management would accept it. It does not seem to have crossed his mind that such union representatives, once identified, might also be quietly fired – as had happened to a number of union supporters at the Teck-Hughes mine some eighteen months before. Mitchell was either wilfully blind or seriously underestimated employer opposition. Employer-dominated committees were acceptable to management, but unions and bargaining were not.

Mitchell's correspondence also conveys his antipathy to this particular union, which he portrayed as hostile, uncompromising, and a user of 'questionable tactics.' His view was similar to King's, who saw the new unionists as 'irresponsible' and not sufficiently 'accommodating.' At no time did Mitchell indicate that support among the miners for Local 240 was a relevant factor in the situation; nor did he acknowledge that the procedure he suggested to McGuire for 'getting one foot in the door'[36] had been attempted by the Cobalt local for some years without ultimate success.[37] There was no indication that the device would be any more successful in Kirkland Lake. By 1941, as a result of its past experience, the union had determined that it must have recognition, collective bargaining, and a master agreement. Anything less was unworkable, given management's intransigence.

The IDIC's role in the Kirkland Lake dispute crystallized the labour movement's dissatisfaction with the functioning of that body. After Kirkland Lake the commission continued to recommend the committee approach (for example, in the Canada Packers' dispute), but by this time the IDIC was entirely discredited. It came to be regarded as merely one more obstacle in the way of union recognition, one more tactic of delay. Moreover, its report in Kirkland Lake convinced other sections in the labour movement of the importance of the issues involved in the Kirkland Lake dispute. The Kirkland Lake Formula appeared to be the latest government policy. This development was to lead to a national response after the conciliation board was created and had reported.

The effect of the IDIC report on management was also significant. The mine operators took IDIC intervention and their proposals as a sign that the federal government was unlikely to force them to bargain – as indeed it did not.[38] They were, therefore, encouraged in their refusal to negotiate with the union.

The IDIC returned to Ottawa and nothing further happened. The Department of Labour was apparently still awaiting the companies' formal response to the union's ten demands, made some weeks before.[39] The government knew the substance of the company position from their complaints about the union's dmands, but presumably it was also considering the IDIC report. The cause of the delay was unknown to the union and was left unexplained.

At two meetings held in the Strand Theatre on Sunday, 10 August, the miners, through their union, had rejected the IDIC recommendations. At the same time the union called a 'holiday' for the purpose of taking a strike vote.[40] This show of strength was to take place on 15 August and the union offered to have the miners work the following Sunday so that production would not suffer. Those companies that had planned production for that Sunday prior to the union's action reluctantly agreed, but the other companies could not rearrange their work schedules and so lost some production.

On the 'holiday,' despite heavy rain, 2000 men picketed the main gates in the pre-shift period. After the whistle blew they went to union headquarters on Prospect Avenue and the voting began. The men cast their votes from eight o'clock in the morning until seven o'clock that night. The picketers were good natured and orderly. The *Northern News*, which was unsympathetic to the union, reported that 'the outstanding discipline of the union members was the outstanding feature of the demonstration, equalled only by their good humour as they stood in the steady downpour for well over an hour.'[41] Some men went through the picket lines. The companies' statement claimed that pickets prevented non-members, as well as union members desiring to work, from entering; however, the office workers were on the job and maintenance men and millmen were performing essential functions since the union had instructed them to report to work.

Tom McGuire was cheered as he drove around to picket lines to tell the miners of their 'brothers' solidarity. At 5:30 PM, in the late afternoon, there was a massive parade of picketers down Government Road. It comprised two divisions: one turned east to man the picket lines in front of the Wright-Hargreaves, Sylvanite, and Toburn mines, and the west column marched to Lake Shore, Teck-Hughes, Kirkland Lake Gold, and Macassa mines. The marchers carried banners that read 'We Want a Conciliation Board,' 'No Company Union Need Apply,' 'Down With Hitler,' 'Increase Gold Production,' 'Trade Unions Did the Job at Kirkland Lake,' 'All for One, One for All,' and 'For Higher Wages.'[42]

The Teck Township police were on hand, but there were no incidents. The provincial police were patrolling other mines in the district, but again there were no incidents. The RCMP were also present.

The demonstration proved that the mines could not operate effectively if the men chose to strike. In one day there was a loss of $22 500 in wages, and the strike curtailed production to the value of $86 845. The strike-vote 'holiday' demonstrated the local union's strength. According to the press, the response of the townspeople was favourable. Four-fifths of them were sympathetic to the union, felt the government had side-stepped the issue in delaying the conciliation board, believed the mine operators had failed to set up any means for employees to deal with management, and believed further that the operators should have made some concessions to their employees.[43]

The companies jointly informed the Department of Labour[44] that this strike vote was an 'illegal walkout' in contravention of the IDI Act, which prohibited work stoppages once an application for a conciliation board had been made. Publicly, the deputy labour minister stated that the vote was illegal. Privately, the department had advised the union, before the one-day 'holiday,' that 'any such vote was unnecessary. But they decided on taking a one-day "holiday" to take the vote notwithstanding the advice of the Labour Department. The vote substantially rejected the acceptance of the report of the [IDIC] Commission.'[45] The strike vote resulted in 2475 votes in favour of strike action and 108 against – a ratio of 24 to 1.[46] The union thus had proved the extent of its representation, and the 'direct effect of this strike vote was to waken Ottawa out of its apathy.'[47] Obviously the problem was not going to disappear.

On 19 August, McGuire, Joe Rankin (Local 240's vice-president), and Bob Carlin were requested to go to Ottawa to confer with the minister of labour and with President Mosher of the CCL. The miners were not impressed by this conference (which merely reiterated the IDIC's suggestions), or by the further delay. Upon his return from Ottawa, Tom McGuire, speaking in Timmins about the Ottawa conference, declared that 'we might just as well not have gone. We held meetings and they tried to ram the company proposals down our throats. They tried to bulldoze us.'[48] He further expressed his view of the IDIC proposals by saying that 'these company union proposals have been brought up to fight the CIO.' McLarty immediately answered this attack without, however, contradicting its substance. He replied: 'There was no attempt to bulldoze anyone. We tried to reason with the delegation since it seemed the mines had gone *a long way* when they agreed to the suggestion of collective bargaining by the mines' committees.'[49] The government was, in fact, inclined to favour the IDIC's recommendations as an acceptable solution to the problem.

5
The conciliation board and the strike vote (August–November 1941)

On 22 August, six weeks after it had been requested, a conciliation board was finally appointed in order 'to consider the proposals of the employees'[1] at Kirkland Lake. The board consisted of the chairman, Mr Justice McTague[2] (McLarty's nominee); F.H. Wilkinson, KC,[3] the employer representative; and J.L. Cohen, KC,[4] the representative for the employees.

Meanwhile the labour dispute continued to dominate the life of the community. On Labour Day, 2000 miners paraded from the union offices along Government Road to the schoolgrounds, where they heard speeches of support. Union President Reid Robinson, who had been detained and held for eight hours by the Department of Immigration as he was travelling to Kirkland Lake, told the miners that the international union would support them. But he still held out hope for a negotiated settlement: '[the companies] talk of a committee of workers from the mines to represent workers in negotiations with them. They have one already. It is called Local 240 and if they want to meet that committee, it can be arranged tomorrow.'[5] Reeve Carter spoke and complimented the workers on their orderly conduct during the taking of their strike vote. The Reverend Father Cavanagh expressed his support and called for the recognition of labour as outlined in the encyclicals of popes Leo XIII and Pius XI. Tom McGuire and Bob Carlin, the union organizers, encouraged the miners, and Jim Russell brought greetings from the Timmins local. At the same time local newspapers in the North became particularly slanderous.[6]

On 6 and 7 October, the board accepted briefs from the parties and conducted public hearings in Kirkland Lake. All twelve mines submitted briefs. The Sylvanite mine's brief, which was presented by K.C. Gray, the mine manager, outlined the general argument of all the mine operators.[7] The other companies agreed in substance with this brief and thereafter proceeded to discuss their own particular economic situations in greater detail. Evidently the Sylvanite brief had been drafted and circulated among the other mine operators for their consideration before they had drafted their own briefs.

COMPANY BRIEFS

The company briefs were largely a repetition and elaboration of the positions that had been taken in the Teck-Hughes dispute and before the IDIC. The companies pointed out that the wages and bonuses were in accordance with PC 7440, the government wage control order, and that, in any event, increased taxes and material costs had narrowed profit margins.[8] The Bidgood mine brief, in particular, supported the wage policy since, by keeping down wages, it allowed the smaller mines to compete on an equal basis for experienced workers.[9] The Upper Canada mine brief noted that its costs were increased because it was located eleven miles outside Kirkland Lake and had to subsidize the transportation costs of its work force, 60 per cent of which lived in the town.[10]

All of the mines stressed the high development costs and the increased cost associated with mining the lower grade ore in response to the government's request for higher production. Further increases in costs could lead to production cutbacks or closure of the mines, for it was 'only by the cooperation of everyone concerned in finding ore and in its profitable extraction that the life of a mine can be maintained.' The Wright-Hargreaves brief was the most definite on this point. It mentioned that to 31 August 1941, it had reserved 500 000 tons of marginal ore, which is interesting as it was the mine most actively making preparations to keep operations going in the event of a strike. However, economic considerations, rather than the state of industrial relations, ruled the mine's policies – notwithstanding its rhetoric to the contrary. Just as it decided that the continuation of operations during the strike was economically advantageous, so it also decided to limit its operation *ultimately* not because of increased wage costs (because the strike was lost and so wages stayed the same), but because of changing world market conditions.

The mine operators (not surprisingly) emphasized their own 'record of courage, initiative, hard work and engineering skill without which we would have no mining industry'; as well as the importance of maintaining an adequate profit, without which there would be no 'desire to create things and attain the possible prize.'[11]

The mines unanimously opposed recognition of the union (which had no right to bargain on behalf of mine employees),[12] and the concept of a master agreement which, it was said, could not adequately deal with the special circumstances of the smaller mines. The union was blamed for decreased production, rapid turnover, and a variety of intimidating tactics. Neither the importance of military service nor the possibility of workers leaving for better jobs in the South was mentioned as a factor in explaining declining productivity.[13] The operators denied the existence of any employer/employee problems: it was the union that was the problem.

The employers had unilaterally introduced a number of medical schemes and welfare plans, and the details of each of these plans were carefully explained.[14] In addition there were provisions for financial assistance, and many of the companies had invested in the town's finance companies and credit agencies. This indirect control over the miners was a source of resentment. Forty per cent of the work force of the Upper Canada mine, for example, lived in a company-run housing project located on the mining property.[15] This situation, common in 'company towns,' meant that many kinds of grievances not normally associated with the employer/employee relationships would inevitably be directed at the mine operators.

Lake Shore mine was the only company that mentioned pensions. It provided monthly pensions to twenty men and contributed a further $7.50 per man per month for pensions. Macassa mine allotted occasional bonuses for efficient work. It also distributed four cash prizes a month of up to $110 for good attendance and safety. The company assured the conciliation board that the company contributed $60 per month towards these prizes 'regardless of what the men put in.'[16]

UNION BRIEF

The union presented an eighty-seven-page brief[17] drafted by Ben Riskin, a close advisor of union president Reid Robinson. Riskin was, by this time, wielding increasing power in the organization and his 'astuteness and intellectual capacity'[18] were undeniable. His involvement demonstrated the interest and concern of the international union. Riskin was assisted by the local union officers and their counsel, J.L. Cohen. The brief's careful drafting, and the depth of its research, impressed the conciliation board and belied the company charges that the union officers were irresponsible agitators.

The union brief identified two major issues in the dispute: wages and the right of the miners to bargain collectively through a union of their own choice. The latter necessarily involved a new definition of management rights, but both concerns were related to the economic and social pressures inherent in the Kirkland Lake situation. The union contended that it represented 75 per cent of the miners and was, therefore, qualified to bargain with the employers. However, despite its majority support, the absence of any legislation requiring the employers to bargain with their employees enabled management to refuse to meet with the union. This was the crux of the union dilemma. The union was ready to meet with employers and denied management accusations that it was 'irresponsible,' but if the conciliation board failed to persuade the companies to bargain, the union would be compelled to strike. The union brief placed responsibility for the breakdown in relations on the mine operators and, by implication, upon the government, which refused to adopt the principles of the Wagner Act.

The union's brief attempted to prove to the board and to the public that there were justifiable grievances that the mine owners refused to acknowledge. Continued management indifference was bad for the workers and the industry, whereas unionization would be positive as it would standardize conditions. 'A policy of more adequate and uniform standardization of wages would assure a maximum degree of cooperation and effort which only satisfied workers can be expected to give to their employers.'[19] The most interesting aspect of the union brief was its detailed account of the living and working conditions of the miners in Kirkland Lake just prior to the strike. To gain 'an impartial estimate' of such conditions, the union had randomly surveyed 1000 miners. In addition, it had done a statistical analysis of prices and rents in Kirkland Lake and Toronto, and then interviewed miners' wives about their living expenditures. The findings of the study were interwoven with the union's demands in an effort to demonstrate that the union's bargaining position had a reasonable foundation.

The wage survey revealed that 83.6 per cent of the workers earned less than the ceiling of $5.20 a day established under the government wage policy PC 7440. The union argued that the board was entitled to consider whether the $4.65 and $5.20 rates were sufficient to meet the level of the high cost of living in Kirkland Lake and the increased cost of living since the outbreak of the war. Management argued that the government-prescribed bonus system was adequate to offset the rise in the cost of living. On the basis of the distribution of annual income earned by the miners, the union concluded that only 16.7 per cent of the employees have 'any appreciable benefit from the contract bonus system.'[20] The wage rates in the different mines were not uniform, and these differences together with 'the illusive bonus' were 'mainly responsible for the labour turnover in the mines.' The union wanted a complete readjustment of wage rates and the bonus system in order to ensure a decent and stable living standard. It contended that this would stabilize employment, and bring about a greater degree of efficiency in the mines.

This union position was not only consistent with the needs of the Kirkland Lake miners, but also with the labour movement's critique of the government wage policy. Given the Peck Rolling Mills decision and others, it is unlikely that the miners' wages could have been raised. There was an indication, however, that the union would have been willing to test the government wage policy[21] had it been successful in achieving collective bargaining. As it turned out, it never got the chance.

The union brief vividly portrayed the housing conditions of the miners. The mushrooming growth of Teck Township had created real estate speculation and escalating prices. 'All building has been undertaken with a view to quick and high returns on investments. Better type of frame houses equipped with modern conveniences and heated apartments are rented at prohibitive prices.'[22] Many miners were building shacks or small houses on the outskirts of the township on property owned by mines. These buildings usually lacked running water, proper sewage, garbage disposal facilities, and other modern conveniences. When concentrated in one section, such settlements created serious sanitation problems that were not conducive to the health and well-being of the community. These squatting areas became known as 'Good Fish Road,' 'Bennett Town,' and 'Federal Town.' Even more substantial frame houses and apartment houses elsewhere in the township had very few conveniences: 'In the centre of the city on Wood Street four apartments on a floor shared one toilet and one bath tub.' Those miners who by sheer perseverance and frugal saving managed to purchase a six-room house usually rented out part of it to carry the mortgage, interest, and maintenance expenses. 'Of the 320 home owners [in the union survey], 19 let rooms and 106 let heated or unheated flats, representing over 30% of the home owning group.' The survey of 700 families (comprising the married group of miners) showed that 'most of the families ... occupy from 3 to 4 rooms, ... only a small percentage rent houses – 15.8% – due mainly to a shortage of low cost houses with a minimum of conveniences. By far the largest number (273) ... occupy unheated flats and apartments.'

Rents varied but most families paid $20 to $30 per month for an unheated four-room house, flat, or apartment. Of the 700 families studied, '109 or 15.55% had no indoor toilet facilities, 231 or 33% had no bathrooms, and 126 or 18% share a bathroom in their quarters with other tenants.'[23] For the vast majority the rent for heated flats and apartments was prohibitive. The cheapest self-contained small apartment cost from $35 to $45 a month. Rented six-room houses, *Labour Gazette* noted, 'are not extensively occupied by working men' and were rented for $35 to $50 a month.[24] The miners in Kirkland Lake did not enjoy a particularly high standard of comfort. This evidence contrasted directly with management rhetoric about the high living standards enjoyed by its workers.

The union brief concluded that the average miner's pay was not adequate to provide a proper and balanced diet for his family because of the general increase in the cost of living during the war, and the unusually high prices of all staple commodities in Northern Ontario. According to Dominion Bureau of Statistics (DBS) figures, the overall rise in the cost of living between August 1939 and August 1941 was 12.8 per cent. The cost of food for a family increased 22.1 per cent in the same period. In comparing the food prices in Kirkland Lake and Toronto, the union found that the average cost of basic commodities was 11 per cent higher in the smaller community. As the lowest priced chain stores did not allow credit buying, the miner often had to go to the higher priced independent stores for his supplies.

From 1939 to 1941 the cost of clothing had risen from 10 to 15 per cent. The cost of various kinds of miner's work clothes had risen from 15 to 48 per cent. The difference in prices between Kirkland Lake and Toronto was insignificant, but miners' families had to dress warmly in the northern climate and needed heavier clothing.

Fuel prices had risen 11.6 per cent from the beginning of the war. Coal of various grades cost from 19 to 40 per cent more in Kirkland Lake than in Toronto. 'In fact, finding coal so prohibitive, many families use wood both for heating and cooking.'[25] Bus transportation in Kirkland Lake was also more expensive. Instalment buying was an important factor in the family budgeting. Of the 700 families surveyed, 53.4 per cent reported instalment payments on household furnishings and equipment. Only 10 per cent reported payments on cars. Automobiles were a luxury that few could afford.

The union brief calculated a minimum budget for a family of five on the basis of the needs of a miner's family, the prevailing prices in Kirkland Lake, and the Toronto Welfare Council's estimate of the minimum needs for a low-income family. The brief concluded that a 'low' budget would amount to $34.71 a week, and a 'high' one would be $40.74:

On the basis of a 15 cent per hour increase, a large group of workers now receiving $4.65 per day would but slightly exceed the requirements of the low cost budget of $34.71 per week, and would receive $35.10 per week, which is still considerably below the more adequate budget of $40.74 per week. Another substantial group of workers in the $5.20 per day category would receive $38.40 per week which is still below the higher cost budget.[26]

For this reason the union demanded a fifteen-cent-an-hour basic wage increase.

The brief also addressed the question of health and safety in the mines: 'a very grave problem which merits serious consideration by the employers as well as by public health agencies.'[27] The mining communities were in constant fear of accidents. Dick Hunter, a representative for the United Steelworkers, recalls these years in Kirkland Lake: 'whenever there was an air blast in town, fear pervaded ...; sometimes 2 to 5 men were killed in one accident. All the women would rush to the mine.'[28] The concern for their safety was one major reason why the miners had turned to the union.

The mining industry had a relatively high mortality rate because of industrial accidents, silicosis, tuberculosis, and other respiratory diseases. Kirkland Lake was no exception and, in fact, had a very bad record. The Ontario Department of Health was also beginning to recognize that lead poisoning posed a serious health hazard for gold miners. According to the Ontario Workmen's Compensation Board, the level of industrial accidents in 1939 was the highest in the mining and explosive industries. This was not an abnormal year. The forty-eight deaths in these industries represented almost one-quarter of all deaths due to industrial accidents in Ontario. Cases of permanent disability and temporary disability were also considerably above the average. Mining was a high-risk industry. The Workmen's Compensation Board reported that the Temiskaming–Cochrane district, a predominantly mining area, reported 2733 cases of temporary disability, 128 cases of permanent disability, and 35 deaths. More specifically, in the Kirkland Lake area, the Kirkland Lake Board of Health in 1940 reported 27 deaths due to accidents and violence, 18.1 per cent of all deaths that year.

Debilitating diseases were another occupational hazard. In 1940 the Ontario Department of Health reported that in the County of Temiskaming the tuberculosis rate exceeded 40 per 100 000 population. In Teck Township there were eight deaths from this disease – a rate of 34.6 per 100 000 as compared to a rate of 24.9 per 100 000 in the population as a whole. The union brief concluded that 'it is evident that the problem of the miners' health is of grave importance, and ... every safeguard should be employed to reduce the ghastly number of accidents and improve conditions which lead to the development and spread of tuberculosis.'[29] The union proposed a better diet, protection against excessive fatigue, proper safety devices, good medical services, adequate hospitalization, and an insurance plan to free the employees from the threat of medical expenses that they could not afford.

The companies had initiated medical plans for their employees prior to the union's arrival. In order to guarantee medical services for the mining employees, mining companies had contracted the services of doctors by guaranteeing they would receive $1.00 per month per employee. This sum was deducted from workers' wages. Employees under this plan received medical attention and a discount on hospitalization charges, but coverage did not extend to miners' families, and did not clearly define the kind of services miners themselves would receive. Consequently, there was serious and growing dissatisfaction with these arrangements. The miners did not feel they were receiving adequate services for which they were paying. In the union survey of 700 miners' families, 391 or 55.8 per cent owed doctor's bills.[30] Most miners had their own family doctors who were not necessarily on contract with the mines.

The mining companies responded to the miners' dissatisfaction by proposing a new medical plan to cover miners' *families* and explicitly outlining the services provided. Such services as existed were broadened from time to time, though they were by no means complete. The cost of this plan was $2.75 a month for married employees, $1.75 a month for single employees. The company involved contributed seventy-five cents per employee per month to the cost of the plan. Despite this substantial employee contribution to the medical schemes all preliminary conferences and planning took place between the companies and a committee of doctors. The employees were excluded.

The miners were suspicious. The plan was to cost them more, but the services had been planned without surveying the actual needs of the community. The employees reacted by requesting (through their union) equal representation with management on a joint committee. The union argued that employee contributions entitled them both to representation and an independent study of the plan before their final acceptance. The union stressed that the miners were anxious to co-operate. Their rejection of the plan was not 'sabotage' as the companies charged, but simply a desire to find a 'suitable, workable, medical arrangement'[31] to ensure good medical services. These services should be administered jointly by the doctors, the companies, and the employees.

The union's concern about accidents and industrial disease, and its critique of the medical plan initiated unilaterally by the companies, raised again the central issue in the strike. The recognition of a union by management meant that such problems would be the subject of bargaining between the parties. The mine operators, however, were unwilling to share their authority in any way, and saw the organization of their employees as a threat to 'take over' the mines. Nevertheless, despite this paranoia, management was correct in realizing that unionization of the gold mines would affect their previously unchallenged managerial prerogative.

After outlining its position on economic issues the union turned to the principal issue in the dispute: recognition. The union argued that 'the growth of the union has been a *natural* development among the workers to meet the need for collective deliberation and action in dealing with various problems affecting their relations with the management of the mines.'[32] The desirability of collective bargaining through an independent union had, the union contended, been supported by PC 2685 and PC 7440.

The press had been especially hostile during this period and that hostility increased during the dispute to the point that the union felt compelled to establish its own paper, *The Northern Citizen*. Union President Reid Robinson, after a particularly vicious editorial, began libel proceedings against the *Northern News*. Despite such attacks the union emphasized 'the unity and sense of responsibility existing among the workers,' and the desirability of establishing collective bargaining as the best means for ensuring orderly industrial relations: 'the public is served both by the stability and continuity of employment so secured, by better earnings and better living standards; and generally by the contribution made to social, industrial life by an industry working in mutual harmony.'[33]

In addition to the general demand for recognition, the union had a number of concrete demands that were to form the basis for further negotiations. The union sought a general 15-per-cent wage increase, and overtime pay after a forty-eight-hour standard (six-day) work week. The hours were to be calculated 'collar to collar' (i.e. from arrival at work to departure), for at that time the miners spent as much as two hours per day travelling to and from their work stations. Travel time was *not* considered working time. The recognition of seniority was proposed as a means of reducing managerial nepotism[34] and the union demanded a formal, enforceable grievance procedure along the lines suggested by PC 2685. Management was, of course, opposed to such procedures as they had demonstrated when their employee committees were established. The union also sought paid holidays (a popular issue in the mines) and supplementary agreements (in addition to the master agreement) with each mine that would take account of the special local circumstances of each of the individual mines. Interestingly (in view of a subsequent letter that the mine operators sent to company shareholders portraying the union as an impediment to post-war re-employment), the union sought an understanding that all miners who had left to join the armed forces would be reinstated to their former employment. Finally, the union demanded a right to participate in establishing a new medical plan 'which would meet all the health needs of the miners and their families [and] is a pressing necessity in this community.'[35]

Reid Robinson assured the conciliation board that the union, a sound organization that had the respect of other large corporations with which it had working relationships, had always administered its contracts responsibly. Robinson outlined the history of the union from its beginnings, and told the board that the IUMMSW (Mine Mill) had 65 000 members and 145 signed contracts that affected the working lives of 85 000 people. Besides its locals in the United States, the union had successfully organized in British Columbia and in the Yukon. One of its largest collective agreements with the American Brass Company (Anaconda) covered 12 000 workers. The master-agreement concept demanded in Kirkland Lake was workable, and was functioning in Utah where the union had a master agreement with fifteen companies. The union administered its contracts and enforced discipline on its members if there were any violations. There was no basis for the employer to charge that unions were irresponsible.

The conciliation board was interested in the union's international and CIO affiliations. These matters had been raised by the mine operators in their brief, and had in recent years been a constant theme in Ontario Premier Mitch Hepburn's attacks upon the labour movement. Robinson explained that the organizational tie reflected the common industrial interests of Canadian and American workers and meant that in the event of a strike the local and the international union would both pay strike benefits from their respective strike funds. He assured the board that the local's dues would remain in Canada, and indicated that in the United States, Mine Mill members paid to the CIO five cents per member per month. The CIO acted as a legislative lobbyist and co-ordinated some of the organizational activities among its affiliated unions. In Canada, such activities were co-ordinated through the CCL, to which Mine Mill's Canadian members paid a per capita rate of two cents per member per month. In the legislative area, the international unions in Canada were autonomous. Robinson compared this structure with that of employer trade associations or chambers of commerce[36] and concluded:

It is true the organization I represent is part of the CIO, but the union of Mine and Mill Workers is an organization which has its own autonomy – that handles its own affairs under the constitution its membership has adopted to govern its affairs and its policies for the future. Its constitution has been adopted by the entire membership of the CIO but the only relationship the CIO can have to the Mine, Mill and Smelter Workers in a situation such as this is that the CIO will support the membership in the event it is necessary to express our indignation at the attitude of the operators; the CIO will not tell us to take such action, but it will back us up if we find it necessary to take such a step.[37]

The international union would back the Kirkland Lake strikers and the union would, in all likelihood, ask the CCL and the CIO for help in the event of a strike.

The companies did not take the conciliation process seriously, and had no intention of changing their position. They apparently felt that the hearing was giving the union a forum and was winning sympathy for its position. Once the presentations of the parties had been completed, the mine operators' lawyer astounded those at the proceeding by declaring that since the union had not justified its case and the companies remained 'unalterably opposed' to any negotiations with Local 240, he was withdrawing from the hearing. This was an unprecedented move. No company had ever unilaterally withdrawn from a conciliation proceeding; but the conciliation board was powerless. If one party remained intransigent there was really nothing that the board could do.

The mine operators later told labour minister McLarty that their lawyer, Mr Lash, had acted 'without their consent or approval. While they had given him a certain measure of latitude, they felt his action ... was both inopportune and unwise.'[38] Nevertheless, the mine operators did not repudiate their lawyer's action, or return to the hearings; nor did they publicly dissociate themselves from his tactics. It seems unlikely that Lash would have undertaken such a dramatic move without the consent of his clients. Indeed, the companies later explained to their stockholders that they had withdrawn in order to shorten the hearing and avoid the possibility of personal attacks: 'The mines believed they had conformed with the letter and the spirit of all peace-time law and wartime regulations and were satisfied to leave their cases for consideration of the board on the merits of the submissions. The mines wished to rest their case on this basis and asked for permission to withdraw.'[39] This was a misrepresentation of what had occurred, but this conduct was not illegal as the law then stood.

The union's careful presentation, Reid Robinson's submission to the board, and the withdrawal of the mine operators from the conciliation hearings gained considerable sympathy for the union's position. Reverend E.E.

Long observed 'a noticeable reaction against the mine operators in public opinion throughout the community.'[40] He recognized that the mine operators' action made it more likely that the conciliation board would favour union recognition, but also realized that the position of the operators would not be easy for them to retract. The two sides were hardening their positions in preparation for the approaching confrontation.

The mine operators' strategy throughout the strike was to deny that collective bargaining was an issue because they remained prepared to bargain with their employees independently of the union. However, the walkout terminated the conciliation board hearings and prevented the board from considering any issue other than recognition: 'by refusing to give evidence, the mine operators themselves make union recognition the only issue on which the board could report and thus chose union recognition as the issue on which to fight.'[41]

THE CONCILIATION BOARD REPORT

On 15 October 1941 the conciliation board unanimously recommended the recognition of Local 240 as the miners' bargaining agent. The board acknowledged that 'what is primarily involved is the right of the workers to bargain collectively through an agent or agency of their own choice with the various employers and arrive at collective agreements.'[42] This was the central issue in the Kirkland Lake dispute. The board report was an implicit repudiation of the IDIC, which had recommended the abandonment of the union and the creation of another bargaining agent more acceptable to the employers. PC 2685 did not authorize employer participation in the selection of a union for collective bargaining or the imposition of conditions concerning the type of bargaining agent with which the employer was prepared to bargain. The board sought support for its position in the language of PC 7440 (the wage control order), which it pointed out 'does not merely declare but orders that all agreements negotiated during the war period shall conform to the principles enunciated in P.C. 2685.' (The government subsequently rejected this interpretation of the wage order and eliminated the confusing clause.) Referring to an earlier dispute in February 1941, the board noted that 'the abstract rights which it is now conceded belong to labour, can only be said to exist in a concrete sense if collective bargaining is practised and collective agreements are concluded. It cannot be said too clearly that labour can no longer be regarded ... merely as a commodity.'[43] While the conciliation board felt compelled 'in all fairness' to recommend union recognition, it was 'under no illusion that the recommendation' was 'likely to be more than a mere formality.' The mine operators' 'technique' had made conciliation futile. The board was powerless to resolve a serious recognition dispute. Because its recommendations were not binding on either party, nothing would come of them.

The board adopted Cohen's argument that the Industrial Disputes Investigation (IDI) Act was an inappropriate mechanism to resolve collective bargaining disputes, and questioned whether recognition disputes were within the purview of the act at all. It concluded that 'the employment of such a technique [management's withdrawal from the hearings] together with the doubt as to jurisdiction under the Act would seem to leave the broad question of collective bargaining to be dealt with by Parliament or Cabinet Council rather than by the old process of conciliation boards under the IDI Act.'

This was a clear challenge to the government. The board had not only repudiated the position of the IDIC, it had also questioned the efficacy of the government's existing labour policy! The report demonstrated that the so-called 'neutral' conciliation process in fact depended largely on the views and personalities of the people making up the conciliation board. J.L. Cohen, the labour nominee, greatly influenced his colleagues, and management's flouting of the board's authority made even the management representative receptive to Cohen's opinions. The conciliation board thus gave the union a propaganda victory, and gave credence to labour's argument that PC 2685 become the mandatory labour policy of the country. Because of its awkward position, the board could not comment on wages and working conditions, which remained matters to be resolved by negotiation 'when the question of recognition has been resolved.' Unlike the IDIC, the conciliation board did not conclude that PC 7440 was the end of the matter. Cohen had, in fact, privately advised the CCL that the Kirkland Lake situation could be used to challenge the wage order.[44]

The other important aspect of the board report was its rejection of the much publicized management position on the CIO and international unions generally. The conciliation board observed that management had never suggested that the union did not represent the majority of the employees. No evidence had been presented to establish that the union was not a proper bargaining agent to represent the employees. 'The assertion was made that it was irresponsible and that the operators were unalterably opposed to bargaining with it, but no evidence of real value was offered in support of the proposition.' On the contrary, the testimony of Reid Robinson had enhanced the union's credibility by emphasizing its successful bargaining relationships elsewhere. The *Toronto Star* subsequently mentioned the same point, and was critical of intransigent employers like the mine operators for their refusal to negotiate with CIO unions. A *Star* editorial observed that 'an effort is still being made to maintain the pretence that by refusing to negotiate with such an [CIO] affiliate, the employing mining concerns will "keep the CIO out of

Canada." The fact is that its affiliates in the Dominion have already a membership of 45 000, that they already hold a number of agreements and that very powerful non-mining interests are now for the first time negotiating with a CIO affiliate for another of these.'[45] The latter reference was to the dispute at the Ford Motor Company in Windsor, where the management had eventually begun discussions with representatives of the United Automobile Workers.

Before the board the mine operators had strenuously objected to the international union link and had argued that such unions should not be permitted to carry on activities in Canada. This was a reassertion of the position – dismissed in labour circles as simply 'anti-union' – adopted by Premier Hepburn during the 1937 strike in Oshawa. It was especially ironic in the Kirkland Lake situation since the union had based its bargaining position on its study of local working conditions, and it was known that the mining companies themselves had American connections.[46] There was no law against joining an international union. Immigration laws had for years encouraged international unions by granting American union organizers free entry into Canada and permitting international labour mobility. The board found that the management view was 'an erroneous and illogical approach to the matter.' This was a timely insertion into their report, for on 5 October 1941, Reid Robinson had been detained at the border by immigration officials (and spent a night in jail) on his way to the conciliation board hearings in Kirkland Lake.

When the conciliation board report was made public, Robinson wrote to Cohen to congratulate him on the 'fine job of drafting' he had done. Robinson believed that the report was a 'constructive move' that would lay the groundwork for a labour-representation law.[47]

Public reaction to the report varied. The *Globe and Mail* as was to be expected, given its connections with the mining industry, opposed the report's recommendation that the union be recognized. Interestingly, it acknowledged that the mine operators had blundered by withdrawing from the hearings. However, it continued to maintain that the CIO was an 'irresponsible' organization that had caused industrial conflict. The *Globe* rejected the American legislative solution: 'we do not want a Wagner Act in this country with the industrial chaos it has brought in the United States.'[48] The *Globe* hoped that the government would ignore the conciliation board report, for 'the country's welfare and the war come ahead of the interests of the CIO treasury.' Various church organizations, however, supported the board's findings.

The conciliation board proceedings and the report gave the local union new confidence, particularly as it estimated that it had 'more than 90% membership of the 12 mines involved.'[49] The membership met on 26 October, accepted the conciliation board report, and set a deadline of 31 October for the operators to respond to union demands.[50] The local met again on 3 November 1941 to request that a strike vote be taken by the labour department as was now required under the recent order-in-council, PC 7307. J.L. Cohen wrote to Reid Robinson that there was 'little likelihood'[51] that the companies would respond to the suggestion of negotiations. Nevertheless, the union's confidence was evident. It even persuaded some local business-men to endorse the recognition of the union. The local also interpreted a recent report of the Temiskaming Presbytery of the United Church as an endorsement of the union's position. A leaflet issued at this time by the union was headed 'The Law, Business, Church and Conciliation Board Back Local 240.'[52]

The mine operators reacted to the conciliation report by circulating their pamphlet, 'The Labour Situation at Kirkland Lake, Ontario – An Important Statement,'[53] to the shareholders. The pamphlet gave the background to the dispute and outlined the economic position of the companies, their position on the matters in dispute, and the submissions that had been made to the IDIC and the conciliation board. The pamphlet was overtly hostile to the 'outside CIO agitators' who had manipulated the 'loyal' employees and thereby induced them to join the union. The language of the pamphlet was less measured than that contained in the briefs to the conciliation board and indicated a real fear of unionization and a strong emotional element in the mine operators' opposition to the union.

The union also distributed brochures and pamphlets. One, 'The Kirkland Lake Situation,' outlined the miners' position in the dispute and responded to the propaganda of the mine operators. Macassa mine immediately fired three men for 'littering' the mine with such literature – an action the union charged was discrimination because of union activity.

CHANGING THE RULES

Following the conciliation board report, the recently proclaimed order-in-council, PC 7307 (published on 16 September 1941), required that a vote be taken in order to determine whether the men would strike to secure the acceptance of the board's report. PC 7307 was proclaimed as a result of an illegal strike at McKinnon Industries in St Catharines involving an inter-pretation of the government's wage control policy. Ultimately the union demand was accepted and work was resumed on the condition that a revision of wages would be negotiated. The government wanted to avoid a recurrence of any similar situation. Strikes could undermine its wage policy. Conse-quently, in accordance with its usual approach to labour policy, an order-in-council was framed that encroached upon the right to strike. It heavily penalized workers for striking *illegally*, while at the same time making it much more difficult to strike *legally*.

This order was the result of a hasty attempt to deal with a particular crisis and as a result was heavily criticized by the labour movement.[54] It was an *ad hoc* measure, neither voted on nor discussed by Parliament, at a time when government by order-in-council was increasingly criticized in both 'progressive' and conservative circles.[55] In addition, the order gave extremely broad powers to the minister of labour. A union was required to notify the minister of its intention to strike or to take a strike vote. If the minister decided that a strike might interfere with the prosecution of the war, he could order that a strike vote be taken under the supervision of his department 'subject to such provisions, conditions, restrictions or stipulations as he may make or impose.' In effect, the minister could determine *who* could vote. The order enfranchised 'all employees who in his [the minister's] opinion are affected by the dispute or whose employment might be affected by the proposed strike.'[56] This left the way open for inclusion in the voting procedure of employees whom a union had no intention of organizing. During the war this usually meant that where a union had attempted to organize the workers in a plant, the government would allow the office workers to cast ballots in the vote which would determine the plant workers' right to strike legally. The only restriction on the minister's discretion was that the vote had to be held no more than five days after the employees notified the department of their desire for a strike vote. The department could not indefinitely delay the issue.

The voting procedure as set out in the order was seen by labour as being undemocratic, for in order that a strike be legal a 'majority' would have to vote in favour; but this 'majority' was defined as a 'majority of those entitled to vote.' The whole weight of inertia was on the side of the employer. M.J. Coldwell pointed out an analogy in Parliament: 'Let me apply a similar test to the Liberal Party. In the last election, 60% of those entitled to vote, voted. The Liberal government received 54% of the votes cast. That is to say they received 34% of the votes of those entitled to vote. Therefore, under their own order in council, if they applied it to themselves, they could not govern this country.'[57] The provisions of PC 7307 were not only compulsory, the penalties for disobedience were particularly high. As labour pointed out, there were no penalties for employers who failed to comply with PC 2685, and no compulsion to do so.

The order was also criticized for contributing to delay: 'Its aim was purely negative. It puts further delays and obstacles in the way of possible strikes, without doing anything whatsoever, to deal with the causes of strikes.'[58] To organized labour, it was one more step in a long legalistic process. Conciliation was a necessary prerequisite before engaging in a legal strike. The IDIC had been made a necessary step prior to the granting of a conciliation board. Now, in addition, the PC 7307 regulations had to be observed. None of these new measures seemed to effectively prevent strikes except in situations like the Teck-Hughes dispute where the delay contributed to the erosion of the union's bargaining position and made the prosecution of a strike impossible.

On 19 September 1941, President A.R. Mosher of the CCL responded officially to the passage of PC 7307:

In passing order in council 7307 ... the government has demonstrated once again its inability or unwillingness to understand the purpose and function of the labour movement and the basic principles which should underlie relationships between employers and workers. Since this order in council imposes still further restrictions upon the rights which have been won by the workers after generations of struggle and sacrifice, without providing any means of removing the causes of industrial unrest, it will have the effect only of increasing discontent and distrust in the minds of the workers, and thus hindering the war effort.[59]

He reiterated his organization's support of the war effort, but emphasized that workers would insist that the democratic principles for which they were fighting abroad must be maintained at home. He advocated government enforcement of PC 2685 and a just wage policy, which, he argued, would reduce industrial unrest to a minimum. There would be no need for 'restrictive and repressive measures such as are contained in this order-in-council.' He also protested that PC 7307 had not been forwarded to the National Labour Supply Council for its consideration: 'labour was deprived ... of an opportunity to express an opinion regarding it before it was enacted.' On behalf of the CCL he protested 'in the strongest possible manner' against 'this ill-conceived and mischievous legislation' and requested its withdrawal. PC 7307 became one more grievance that contributed to the deteriorating relationship between government and the labour movement.

The strike vote in Kirkland Lake was the *first* to be taken under the new order. The manner in which the vote was conducted disturbed both the union and the employer and increased the union's suspicion of the government's motives.

On 3 November 1941, the day Local 240 requested a government-supervised strike vote, two department of labour officials, F.E. Harrison and M.J. McCullagh, arrived in Kirkland Lake. The next day these officials, on instructions from the minister of labour, tried to arrange a meeting of both parties in an effort to mediate the dispute. When the mine operators refused to meet, the officials proceeded with arrangements for taking a strike vote. On 4 November, the suggested rules (as drafted by Harrison and McCullagh) were submitted to Alex Harris, secretary of the Mine Owners' Association. Later that day they were discussed with union representatives. The union regarded these rules as 'fair.' The employees eligible to vote were those who had applied for the conciliation board. Shift bosses, captains, technicians, office staff, and managers were explicitly excluded. Copies of voter lists were to be made available to each of the parties to the dispute. After some discussion there was agreement that the proposed ballot would read: Do you favour strike action to press for the Conciliation Board report? This wording was recommended to Ottawa. There were to be three scrutineers: a returning officer, a representative for the companies, and a representative of the union.[60] Management's position was unclear about these proposals. In Kirkland Lake they agreed to them. At the same time a delegation of mine managers (including representatives from the Toburn, Lake Shore, and Sylvanite mines) were lobbying in Ottawa for a change of rules that would be more to their advantage.

Initially, it seems they had some success. A committee of three had been appointed by the labour department. One member was Humphrey Mitchell. On 6 November, when the department finally responded to the proposed vote regulations, the rules had been changed considerably. Voter lists were not given to the union until one-half hour before the vote began on Saturday, 8 November, thereby making it more difficult to mobilize support and overcome the effect of a voting system that made a failure to cast a ballot the equivalent of a 'no' vote. No list for Golden Gate mine was ever received. The new rules were given orally to the union; the department refused to provide a written copy. The union saw a printed statement of the rules for the first time on 7 November, when they were made public. Yet the *Northern Miner* had this information before Local 240. The voting regulations were printed in the 6 November edition (which was available in Toronto on the afternoon of 5 November), 'at least 28 hours before being presented to the union.'[61] It seems likely that the information was 'leaked' by the Department of Labour. In any event, the mine managers knew the new rules twenty-four hours before the union, and the information was spread throughout the mines on 5 November; it was disseminated at employees' council meetings held that evening. Those eligible to vote were to include *everyone on the payroll* except the president, management, or directors of the company. This was McLarty's interpretation of the phrase 'those affected by the dispute' that appeared in PC 7307. So long as the union had to achieve a majority of those 'eligible to vote,' the definition of the voting constituency remained a prime concern. If the voting constituency were extended to include large groups of employees who had no direct community of interest with the miners and whom the union had not sought to organize, it became much less likely that a vote would authorize strike action. The right to strike for collective bargaining could well depend upon the vote (or failure to vote) of those who had previously expressed no interest in bargaining collectively.

Under the 'new' rules, the scrutineers would consist of a three-man committee at each mine: one government representative, one representative of the employers, and one representative of the *employees* (not necessarily of the union). The union insisted on a written understanding from the department officials that the union would name the scrutineers at each mine and that the committees counting votes would include a union nominee. On 7 November these new regulations were announced by the government. An impartial observer commented that considerable bitterness was created in Kirkland Lake because of an 'obvious leak in departmental information' and because 'McLarty had said one thing and done another.'[62] The union had expected that only miners and surface labourers would be able to vote, as these were the workers the union sought to represent. These men composed what would be termed in later years an 'appropriate bargaining unit.' Their working conditions created a community of interest making it appropriate that they bargain together as a distinct group. Had the Wagner Act system been in effect, such problems would not have emerged.

Immediately upon receipt of the draft of the *new* rules, the union began to exert pressure on the government in an effort to force a change.[63] Trade unionists across the country registered their support. The labour movement opposed PC 7307, and deplored both the government's interpretation of the order and its last-minute changes in the rules. Management also pressured the government in an effort to maintain the *status quo*. An indication of the union's concern was published in a pamphlet that read: 'Honourable (spare the word) Norman McLarty has adopted German methods to double-cross the workers ... We'll let them throw McLarty in [the vote] and still lick the hell out of them.'[64] A more moderate tone was taken by the recording secretary of Local 240, Larry Sefton, who on his regular radio broadcast described the union's campaign to force the government to change the rules. He informed his audience of a telephone conference, the first of its kind in the North, that had been held across Canada and the United States: Norman Dowd (executive secretary of the CCL in Ottawa), Charlie Millard (executive director, SWOC in Toronto), Sol Spivak (director, ACWA in Toronto), J.L.

Cohen, KC (counsel for Local 240, Toronto), Robert Livett (president, District no. 18, UMWA, Calgary), Tom McGuire (IUMMSW, Kirkland Lake), Reid Robinson (international president, IUMMSW, Washington, DC), and M.M. MacLean (secretary-treasurer, CBRE, Ottawa) had all participated and had 'discussed the whole undemocratic pattern.'

In the hours immediately preceding the vote, messages from across Canada urged McLarty to reject the latest set of 'rules' and return to those previously agreed upon. Two fairly representative protests came from the United Mineworkers of America, the other large mining union. Robert Livett wired McLarty protesting the refusal of the government to allow the union (as distinct from the employees) to choose the scrutineers in the forthcoming strike ballot, and asserting the union's right to the voter lists. He considered the government's methods 'totalitarian' and 'unacceptable.' Despite his union's expressed policy of maintaining industrial peace, he declared that industrial peace 'cannot in future be guaranteed unless labour is given a square deal and your government follows a policy of recognizing and enforcing the principles of collective bargaining which your government enunciated in P.C. 2685 in the Kirkland Lake area and applying to all trade unions under similar conditions throughout Canada.'[65] Silby Barrett of the UMWA alleged that the government's inclusion of officials and mining staff among those eligible to vote was contrary to a previous understanding between union leaders and government officials. He concluded, 'You are going out of your way to give aid and comfort to the mining interests in that field.'[66] Charlie Millard declared that he would personally support any action necessary to ensure justice. As Sefton told his radio audience, 'In the face of this barrage of labour opposition, McLarty backed down.'[67]

There was additional pressure put on the minister. A delegation from Local 240, composed of William Simpson and Bob Carlin, went to Ottawa. Along with the CCL president, A.R. Mosher, they met with the minister of labour and strongly urged a change in the voting regulations.

Two CCF MPs, Clarie Gillis and Angus MacInnis, raised the matter in the House of Commons. Angus MacInnis, as acting CCF House Leader, tried to provoke a debate on the issue by asking that the Kirkland Lake strike vote be considered a matter of urgent public importance. He contended that the issue was urgent because the vote was to be held the next day and 'the application of the provisions of this order which has been proposed by the minister in this case will, if carried out, further intensify industrial unrest in this industry and threaten industrial peace throughout the country, with consequent harm to our war effort.'[68] Prime Minister Mackenzie King replied that the vote could not be considered urgent enough to require immediate discussion since it had yet to be taken. The minister of labour was simply carrying out the law of the land. Later, when the House was in committee, Clarie Gillis was able to ask McLarty how the voting constituency was to be defined. McLarty had just concluded a meeting with the delegation from Local 240 and had appointed a three-man committee from the department to make regulations that would be fair to *both* parties. Since this committee was reconvening that afternoon, the minister indicated that he could not answer Gillis' question.

The mine operators were also pressuring the government and suggested that they might have to close the mines indefinitely if there was a strike. On 7 November 1941, a front page story in the *Globe and Mail* reported that the government would take no action should the mine operators decide to close their mines. Dr Mutchmor of the United Church immediately wrote to McLarty about this 'disturbing' article. As he saw it, 'such a threat from the mine operators should not go unheeded; neither should it be allowed to intimidate men in casting their ballots on the fundamental issue of collective bargaining.'[69]

To gain sympathy for their position, the mine owners professed a deep sense of obligation to the 1200 men who had left the mines and gone to the war and had been promised re-employment upon their return. This argument was the basis of an appeal for shareholder support: 'We ask you to consider how insecure we would feel as to our ability to fulfil what must be termed a sacred trust to these men, if we, by recognizing the CIO union, place the destiny of these returned soldiers and airmen in the hands of a body which has not only proved itself irresponsible and untrustworthy but ... has revealed a thoroughly anti-British sentiment. This we refuse to do.'[70] The anti-British sentiment referred to in the letter was an attack on the American connection and Reid Robinson's acceptance of American neutrality. Following the American entry into the war, the union could no longer be characterized as anti-British; however, the companies continued to question the loyalty of the striking miners and to alarm the shareholders with suggestions that the CIO planned to take over the industry.

The strike vote was scheduled to take place on 8 November. On the evening of 7 November Tom McGuire was advised by telephone of a further revision of the Department of Labour's voting regulations. The department had reversed itself again and now agreed to exclude all supervisory and technical staff (including engineers), office staff, management, the presidents, and boards of directors. Union handbills immediately announced: 'Labour

Forces Government's Hand.'[71] But later that evening the union was advised of the 'revised' wording of the question on the ballot. The copy they were given

was typed on a Remington typewriter ... It was mimeographed on ordinary mimeo-graph paper, was completely devoid of any numbering system whatsoever, and read 'Your employer ... agrees to negotiate with a Committee elected by the employees of the Company but not with Local 240, of the Mine Mill and Smelter Workers Union.

Are you in favour of a strike unless	YES ___
the company is prepared to negotiate	
with Local 240?	NO ___

Vote by marking X.'[72]

This was a revival of the employee-committee idea that the IDIC had pro-moted some three months before, and which the union had firmly rejected. Humphrey Mitchell had, of course, both participated in devising the voting regulations and chaired the IDIC that advocated the committee solution. The union consulted Cohen, but it was decided that it was too late for anything to be done. Voting was to begin at 6:00 AM.

THE VOTE

Nine men in addition to McCullagh and Harrison came to Kirkland Lake to conduct the vote. The voting proceeded peacefully. There was a 'surface calmness'[73] in Kirkland Lake throughout the day. Despite the absence of turmoil, the conduct of the vote generated criticism. Voter lists were given to the scrutineers immediately prior to the vote. The lists were not given to the union beforehand and there was no opportunity to check their accuracy. The procedure was inefficient and the rules inconsistently applied. For example, surveyors were excluded at one mine and prospectors in Quebec were included on the eligible list at another. These prospectors would not cast a ballot, and thus would count as 'no' votes. Compensation cases were ruled out at the Upper Canada mine, but similar cases and the inmates of an insane asylum were counted in at others. Supervisory staff were allowed to vote at some mines. Cooks were generally ruled eligible, in direct contravention of the promises made earlier to the CCL executive. The union viewed the gov-ernment's role in the vote as obstructive and believed that 'treachery and inefficiency were the most obvious characteristics of the Department of Labour conduct all through the conduct of this vote under order in council P.C. 7307.'[74]

Clarie Gillis was in Kirkland Lake to observe the vote. Later, in the House of Commons, he attacked the voting procedure, the failure to provide the union with a voter list, and the wording of the ballot with its implicit appeal for company unionism. He warned the House: 'There is now developing in the Kirkland Lake area a situation which if allowed to continue, will have serious repercussions not only in that area but perhaps all across Canada and will eventually ramify into the Department of Munitions and Supply because the same situation is developing at Ford Motor Company in Windsor where the machines are made that are necessary to our war effort ...' He concluded his address by predicting (accurately as it turned out), that 'if a strike takes place, it is going to affect the entire labour movement in Canada, because there has been so much injustice and disregard of regular procedures in that area that the entire labour movement in Canada will take it up, and perhaps we shall have trouble from coast to coast ...'[75]

Supporters of the mine operators were also dissatisfied with the determination of the voting constituency. There were protests to McLarty from the disfranchised workers.[76] Managers W.K. Summerhayes and K. Gray of the Wright-Hargreaves and Sylvanite mines, respectively, protested the last-minute changes of the voter list. Gray complained that at his mine 41.8 per cent of the total work force were 'foreigners' and more than half of those were not naturalized. Such people had been allowed to vote, whereas his staff, which was '100% Canadian,' was excluded.[77]

The *Northern Miner* attacked what it termed 'the bungling attitude' of the Department of Labour and warned that if a strike resulted, the government deserved a portion of the blame. Headlines read, 'Enemy Aliens Play Part in Crippling of War Industry,' 'Kirkland Lake on Brink of Strike.'[78] It was shocking, reported the newspaper, that unnaturalized aliens – including 'many enemy aliens' – were allowed to vote when many British citizens were disfranchised. The report greatly exaggerated the number of immigrants who had voted. One suspects that the real concern was that over 500 employees who could have swung the vote against the union had lost the right to vote. The newspaper noted, 'These were salaried, technical and other workers and the vast majority were against the strike.' The ethnic origin of some of the employees was not a dominant factor in explaining the union's support, nor was there any 'plot' by enemy aliens. In view of the way in which the question on the ballot was framed, the results of the vote demonstrated unequivocally that the employees supported their union and not a management or government-sponsored employee committee. The ethnic references in the newspaper account were included for their emotional appeal in the context of the war, and were meant to minimize the union's victory.

The Toronto *Financial Post* was among the most vitriolic of the newspapers. It, too, pointed out that Kirkland Lake had a huge alien population that included 'many enemy aliens.' These people had been allowed to vote while 460 'British' had been disfranchised. The government was allowing 'the CIO and its fellow travellers [to] muscle into Canada's mining industry.'[79] *Globe and Mail* editorials took the same line.

The results of the vote indicated that the majority in eight of the twelve mines favoured strike action. In two of the smaller mines – Golden Gate and Upper Canada – the vote was insufficient to support a strike. The Brock mine had already suspended operations and Morris Kirkland was bankrupt at the time of the vote. Of the 4333 eligible voters in all the mines, 4057 voted. Of these, 2725 voted to strike and 1254 voted not to strike.[80] The union was supported by 63 per cent of those eligible to vote, and it received 67 per cent of the votes cast. Some 31 per cent opposed strike action and there were a few spoiled ballots. The overall majority in the twelve mines favoured a strike, but a strike was legal only in those eight mines where the union had received the support of a majority of those eligible to vote. In two other mines, the union received a majority of those who voted but not of those *eligible* to vote.[81] The *Canadian Forum* compared the Kirkland Lake vote with one taken in a number of Canada Packers plants in Toronto. That vote involved a contest between a union and the plant committee that the union had won in several plants, although it had not achieved an overall majority. In that case the government had determined that the relevant figure was the union's support 'across the board.' Accordingly the union was not recognized even in those plants where it achieved a majority. In Kirkland Lake, on the other hand, the union *had* achieved an overall majority but was not allowed to strike all of the mines. The government decided that the union's majority in Kirkland Lake must be calculated for each mine. As the magazine commented, 'evidently it all depends who gets the *majority*.'[82]

The strike vote was the climax of the union's organizing drive, and brought the strike one step closer. The membership had 'held' and were determined to support the recognition of their union. Just prior to the strike, Tom McGuire wrote to J.L. Cohen, and estimated that by October 1941, well over 3500 miners had paid union dues, including 250 new members who had just joined the union. In the three locals in the area there were approximately 5000 members. Despite the operators' attempts to establish employees' committees and the circulation of anti-union letters to the miners and their wives, McGuire reported that 'we are in the strongest position ever in the history of the organization.'[83]

The *Northern Miner* commented upon the important psychological impact of the strike vote and the government reversal: 'Immediately on receipt of the change, union agitators revived their line and cry that the government was susceptible to coercion and that a vote for the strike was not a vote to quit work but only another step in the program to force union recognition.'[84] The paper characterized the dispute as one between the CIO and the government, and claimed that the CIO had been able to capitalize on the change in the government's rules that gave the impression that the government was weakening. The paper asserted that 'the vote was not truly a vote on the issues in dispute but was the development of a theory that the government could be intimidated.' Since the government had played into the CIO's hands it would be responsible if CIO 'disaffection' spread throughout the entire mining industry.

This view was an exaggeration. Undoubtedly the government's reversion to the voting rules that had originally been agreed upon by the parties encouraged the union. However, in view of the government-imposed delay before they could be in a legal strike position, the miners were under no illusions about the difficulty of their struggle. They were aware of their employers' adamant resistance to a union. They had had months to make their preparations for the strike. The strike leaders were active, idealistic men, but they were not fools. Prior to the strike Bob Carlin had expressed doubts that the union was strong enough to combat the combined business and government interests that opposed the union. Because of the long delay, the strike would be fought in the winter. Carlin recalls: 'I told McGuire ... I think we're going to lose the strike now ... I had got the idea it had drifted along ... We had already taken a strike vote – we had two strike votes – one conducted by ourselves and one conducted by the government, and yet we weren't in bargaining and were coming close to the fall ... and I was in a strike where it was lost because of Old Man Weather – the winter you see.'[85] This prospect disturbed the union leadership. The past history of strikes in the mining industry was not encouraging: 'Anytime in June or July we could have sat down with the operators ... but they had dragged us into November ... They knew that we couldn't beat Old Man Winter ... because they'd beat us on four or five occasions before in Cobalt, always in the winter – the summer we won and in the winter they won.'[86]

As early as the previous October the strike leaders and the CCL executive board had discussed the feasibility of strike action and had determined it would be risky. A great deal depended on the federal government's willingness to intervene in support of the principles of PC 2685 and impose union recognition on the mine operators. It was becoming increasingly clear by the fall of 1941 that the government had no such intention. The union was not optimistic, although it did not wish to discourage its supporters, who had after all risked their jobs to demonstrate their convictions. The employees wanted union recognition and collective bargaining, and the union was determined to mobilize organized labour in support of that objective.

6
Broader support for the impending strike and the final mediation effort

In the weeks prior to the strike, both sides mobilized their supporters and intensified their publicity campaigns. On 4 November, immediately prior to the strike vote, the issues were debated in the Teck Township Council. A resolution, sponsored by Reeve Carter and Councilman Mickey Maguire,[1] urged the Dominion Government to adopt legislation requiring compulsory recognition of any union freely selected by the majority of employees. After the strike vote the debate intensified, since the strike appeared to be unavoidable unless one party would yield. Compromise seemed impossible. The operators would neither recognize nor meet with the union, and the union could not acquiesce in a position that denied its existence and legitimacy. The operators publicly offered to negotiate with 'employees' committees' – the mechanism suggested by the IDIC (Industrial Disputes Inquiry Commission) – but the *Northern News* reported that they were finalizing their preparations for the strike.[2]

As the collective-bargaining issue in Kirkland Lake gained prominence, support from outside the community increased. The union played host to numerous trade union leaders and politicians. Dorise Nielson, 'Unity' MP for North Battleford,[3] addressed the Kirkland Lake women at a meeting in the Strand Theatre, sponsored by the ladies auxiliary of the Union: 'The eyes of all Canada are on Kirkland Lake. The whole future of organized labour in this country depends on the battle you are waging here ... This may turn out to be the greatest labour battle in the history of Canada.'[4] She urged the women to support a strike if it became necessary.

The mine owners were also mobilizing. They informed their 75 000 shareholders[5] about the situation in Kirkland Lake and assured them of their opposition to the union. On 10 November, at the annual meeting of Wright-Hargreaves mine, the shareholders enthusiastically endorsed the mine operators' policy.[6] On 18 November the Ontario Prospectors and Developers Association wired the prime minister, the labour minister (Norman McLarty), and the minister of mines (Mr Crerar) of their support of the mine operators' position. The association believed that the recognition of a foreign-controlled union would impede the prosecution of the war[7] and, ignoring the foreign ownership of some of the mining companies, argued that 'control' of the industry by a foreign union would discourage prospecting.

During the course of the Kirkland Lake dispute, the Canadian Congress of Labour (CCL) had also become increasingly involved, providing at first rhetorical, and later more tangible support, as it became apparent that the struggle involved issues of general concern to the entire labour movement. As early as October 1940, the journal of the CCL, *The Canadian Unionist*, had discussed the inadequacies of PC 2685,[8] but particularly the failure of unionized workers to secure employer recognition. It pointed to the Teck-Hughes case as an example that had 'brought matters to a head,' and criticized the mine owners' 'point-blank refusal to accept any part of the [conciliation board's] recommendations,' including those concerning union recognition.

By early 1941 the congress had become alarmed by the government's failure to enforce PC 2685, and the resulting adverse effect on trade union organizations. This issue was a major point in the CCL's annual memorandum presented to the government in February 1941. Bitter recognition disputes, like the one at National Steel Car, were occurring more frequently; and labour's reservations about the wage control policy increased as rigid interpretations restricted the scope of collective bargaining.[9] In the absence of legislative recognition, protection, and support, employer opposition and the restrictive wage policy could indefinitely stifle the growth of the labour movement. This general dissatisfaction and the deterioration in government/labour relations focused even more attention on the Kirkland Lake dispute. As a result, a number of senior CCL officials became personally involved.

In March 1941, Aaron Mosher, president of the CCL, addressed the Kirkland Lake community over radio station CJKL. He struck a moderate tone, and tried to counteract the charges of foreign domination by stressing the autonomy of both the union and the CCL. He assured his listeners that neither organization was controlled by the CIO and that collective bargaining was a positive force that would benefit workers who were being treated 'like so much raw material.'[10] These remarks were unlikely to change many opinions, but they were good publicity for the union.

In the summer of 1941 (i.e. when the union organizing campaign was peaking and the IDIC was asserting that employee committees were compatible with the principles of PC 2685), the CCL again addressed the inadequacies of the government's labour policy, and the failure to support the principles of PC 2685: 'Since the workers were being asked by the government to give up during the war their right to obtain higher wages, except where the wage levels could be shown to be "unduly low or subnormal" and to accept a cost of living bonus instead ... the government should require employers to accept the principles of P.C. 2685 with regard to union recognition and collective bargaining.'[11] The government had given widespread publicity to its wage policy, but no such publicity to its labour policy. It had not even applied its labour policy in government-owned and controlled plants.

In August 1941, Norman Dowd of the CCL spoke to the Canadian Institute of Public Affairs and criticized the government's position as evidenced by its conduct in several disputes, including those at National Steel Car and Peck Rolling Mills. He pointed out that after a dispute at Chrysler, an order-in-council was passed restricting picketing. The Arvida dispute in July 1941 had resulted in an order-in-council allowing the federal government to by-pass consultation with provincial and municipal authorities and unilaterally to call out the militia in a strike situation. Dowd also pointed out that a number of trade unionists had been interned for organizing unions and had been deprived of their civil liberties. 'These are just a few of the things that have caused deep concern in the minds of many workers.'[12]

In September 1941, A.R. Mosher addressed the CCL convention along similar lines. The government had not given proper consideration to the labour movement. Several disputes had been aggravated by 'dilatory and unwise action.' The government's failure to treat workers fairly set a bad example to employers who were acting in a similar manner. In such circumstances, the CCL had to adopt a policy that would change the government's attitude and help to overcome the opposition of employers. What was needed was industrial organization.

The most important factor in accomplishing both these purposes is the development of economic and political power by a wide extension of organization in every field of industry. To the extent that workers become organized, they can force the employers to recognize their unions and negotiate agreements with them and at the present time this is the course which the workers are following, subject to the limitations imposed upon them by the IDI Act.'[13]

In addition, labour had to continue to pressure the government to protect the right to organize and bargain collectively. In the same month Mosher attacked PC 7307, which he considered a denial of the right to strike.

In October 1941, the CCL expressed considerable interest in the conciliation board proceedings in Kirkland Lake and offered every assistance in bringing the dispute to a satisfactory conclusion. Upon hearing of management's withdrawal from the board's hearings, Mosher wrote to Tom McGuire and advised that the CCL would support 'to the fullest extent of its ability' the union's efforts to get a settlement.[14] The extent of the congress' commitment had yet to be defined.

The conciliation board report on the Kirkland Lake dispute was a turning point. Prior to the report there was interest in the dispute and an awareness that it was important; but after the report had highlighted the issues of union recognition, the inadequacy of the existing industrial relations machinery, and the intransigence of management, the labour movement undertook to formally support the strike. The *Canadian Unionist* reported that 'the issue of collective bargaining with the union of the workers' choice is now more prominently before the public than ever, as a result of the unanimous report of the conciliation board in the Kirkland Lake gold mines dispute.'[15] The government's behaviour was a catalyst in unifying labour behind the Kirkland Lake miners and reinforcing their determination to take industrial and political action. It was clear that the mine operators would not accept the conciliation report, despite the fact that it was unanimous and favoured by the company nominee on the panel. There would be a fight and some decision by the CCL had to be taken.

This CCL decision was made at an executive committee meeting on 21 October 1941 and endorsed by the executive council the following day. At these meetings the Kirkland Lake situation was given 'very serious consideration,'[16] but what is fascinating about the discussion is the complex motivations of the men who led the new industrial union movement. Their discussions reflected many elements of both narrow 'business' unionism, which emphasized bread-and-butter issues, and a broader 'social' unionism, which took account of the social and political aims of the labour movement. The improvement of people's living standards was important; but so was the drive for 'industrial democracy' – the right of employees to participate through their own organizations in the decision-making process of business and government. The Kirkland Lake strike was both a struggle to better working conditions in Kirkland Lake and a fight for the principle of collective bargaining. It would educate the public about unions and the miners themselves about the value of solidarity. It was also a political battle to achieve from the government the necessary legislative protection for unions and collective bargaining. Since previous lobbying had secured only unenforceable declarations, it seemed necessary to rely on economic power, which the gov-

ernment could not ignore. This was a collective demand by the new industrial union movement for status – for the recognition of workers' rights on
the shop floor, and for the recognition of these new industrial unions themselves. It was a demand for involvement through bargaining in the decision-
making process in industry, and for involvement in government through
representation on committees and direct consultation with policy makers.

The meeting began with a report from Tom McGuire and J.L. Cohen
outlining the extent of local organization, the efforts of the operators to
organize employee committees, and the prediction that the union would win
a strike vote. On the previous day the conciliation board report had been
presented to the minister of labour, so Cohen outlined management's over-
confident approach to the conciliation board and explained that the essential
issue was the opposition to dealing with Local 240. The mine operators had
not tried to address the issue of the high cost of living in Kirkland Lake, but
had merely argued that they were operating in accordance with the government's wage policy. Cohen explained that the conciliation board report highlighted the collective-bargaining issue, and dealt little with wages. The
union's right to engage in collective bargaining was the only serious issue
and the mines might be willing to close down rather than deal with the local.
He did not mention the very real possibility that the mines might try to
continue to operate.

Pat Conroy of the CCL and J.L. Cohen both believed that the issue was
one of *national* importance and urged the congress to provide financial support: 'We are fighting for collective bargaining; we have taken issue with the
government on that point.'[17] Mosher was more cautious and conservative.
He took what he considered to be a 'realistic' view of the situation, and
sought Cohen's opinion on the effect of the new wage legislation on the
dispute. Could wages be made an issue? Mosher was worried that if union
recognition were the only issue there would be no room for compromise.
Victory or defeat would be total.

McGuire promised that the union would seek support for its Ten-Point
Program, although he believed that it *could* mobilize support around the
issue of collective bargaining. Mosher was more sceptical. Recognition was a
sophisticated issue that the public would not understand. As trade unionists
they knew that *in practice* employee committees became the creatures of
management; but the public, with no experience in collective bargaining,
would not know this. The offer of 'democratically elected' employee committees would *appear* to be a reasonable compromise, even though historically such committees had been a facade designed to thwart legitimate trade
unions. The *reality* of employee committees was the antithesis of a freely

selected bargaining agency, but few citizens outside the labour movement itself were acquainted with that reality.

Mosher doubted that the congress could raise money from its affiliates to support the Kirkland Lake strike, and further doubted that the local union members in Kirkland Lake would support a long recognition strike: 'We won't convince 5% of [the] masses or 5% of [the] workers – not 10% of [the] workers would assess themselves $1.00 a month for collective bargaining. If the International Union hasn't the money and is depending on Canadian unions – we are leading men into fool's paradise. To be brutally frank, I don't like it.'[18] Conroy was aware of the broader implications of the strike, and believed that the congress should solicit the support of its affiliates. Mosher, on the other hand, drew from his own experience and recalled how difficult it had been to persuade the Canadian Brotherhood of Railway Employees (CBRE) members to support a strike for the *principle* of collective bargaining.

Charlie Millard had been president of the United Automobile Workers (UAW) local in Oshawa in 1937 but attended this meeting as head of the Steelworkers' Organizing Committee (SWOC) in Canada. He supported Conroy and praised 'the readiness of people to pioneer and to fight for principles.' Millard was convinced that the labour movement must press the issue of collective bargaining, though in his own mind he was still wondering whether Kirkland Lake was *the* appropriate strike.

Sol Spivak, leader of the Amalgamated Clothing Workers of America (ACWA), entered the debate to conciliate the conflicting views of Conroy and Mosher. Mosher took a conservative, pragmatic view that would have limited the extent of CCL involvement, while Conroy wanted to make the strike the CCL's fight for collective bargaining. Spivak pointed out that many strikes had been lost in recent years and that even the strongest union could not guarantee a victory. He agreed that the congress should support the Kirkland Lake strike by doing what was *possible*, but he warned that the miners could not look to the CCL to answer all their needs. Despite these different views it became clear that the congress would have to support the miners' union along the lines Conroy had outlined, for the issue was fundamental to the new union movement.

What kind of support would be given by the CCL? What did the IUMMSW (Mine Mill) want? Cohen reported that the international union would provide reasonably adequate financial assistance, but that the amount it was willing to commit to the strike would depend, in part, upon the extent of Canadian support. Conroy estimated that the strike would cost $25 000 a week, and Mosher believed the union would need at least half a million to win the strike. Mosher suggested that Reid Robinson seek CIO support, but it was recognized that such support was difficult to obtain because the CIO felt it was no use spending money in Canada unless a legislative basis for collective bargaining was established.[19]

Ultimately the executive committee decided that in the event of a strike in Kirkland Lake, the CCL's affiliated unions would be asked to support the strikers. In the meantime, the congress would adopt its traditional role of pressuring the government about PC 2685 which, if enforced, would make the Kirkland Lake strike unnecessary. In this regard it was decided that the TLC should be approached so that both congresses could co-ordinate their attack on the government. Conroy was pleased that the CCL was receptive, but he still believed that it was desirable to avoid a strike if at all possible.

The decision of the executive committee had to be approved by the larger gathering of union representatives in the executive council on the following day. At this meeting, further discussion of the congress' *financial* commitment to the strike took place and the same reservations were expressed. The members of the executive committee, although they were heads of unions, could not give any assurances as to the extent of their unions' support. Mosher informed the council of the estimated cost of the strike and of his own doubts about the CCL's ability to raise money: 'it is an important issue and we are prepared to stand behind them [the miners] ... [but] no great volume of financial support can be obtained.' The miners must realize that fact: 'I wouldn't want them to feel that even 1/5 of what is required for a week could be secured by the Congress.'[20] Conroy was more optimistic about the amount that would be forthcoming. The CCL *could* publicize the strike and mobilize support among its affiliates.

Tom McGuire and Charlie Millard estimated that the struggle would be a long one. The four largest producers were involved, some sectors of the business community would support the mine operators, and the government was not interested in settling recognition disputes, as had been demonstrated in the National Steel Car, MacKinnon Industries, and Nova Scotia coal strikes.[21] Nevertheless, the council approved the executive committee's decision and

unanimously decided that these executive bodies would recommend to Congress unions that full support be given to the mine workers in their struggle for the right to have the union of their choice recognized and an agreement negotiated with it. The issue involves the labour policy of the government and it can scarcely fail to take action in the face of a unanimous Conciliation Board report which is based unequivocally on principles enunciated and advocated by the government.[22]

That same day, the CCL executive met with the minister of labour (McLarty), accompanied by a delegation from its Kirkland Lake local. They stressed the necessity of making the principles of PC 2685 mandatory, but McLarty only undertook to place the views of the congress before the government.

The CCL executive recognized that the strike was a crucial one for the future of organized labour, but did not make *direct* financial contributions, or take extraordinary measures to mobilize support until well after the strike had begun. The primary functions of the CCL were legislative and political. Except with respect to its directly chartered locals, it had no role in organizing and collective bargaining. These functions remained the prerogative of the affiliated unions, which did not encourage an independent role for the congress in such matters. Any assistance by the CCL was directed and controlled by the leaders of the affiliated unions through their membership on the congress executive. As a result, the central labour body remained dependent upon its affiliates. In any event the CCL was a relatively new organization with limited resources.

Following the CCL decision of support, a bulletin was mailed out to all of the secretaries of affiliated unions and all of the labour councils to inform them of the position taken by the CCL executive and council. The bulletin was to be read at all union meetings, provided background information for union members, and stressed the 'importance of giving [the miners] the strongest possible support in the event that a showdown is necessary.'

The day before the strike began, a second bulletin was sent at Tom McGuire's suggestion, recommending that all members write the prime minister and their MPs to urge that the mine operators be compelled to accept the unanimous conciliation board report. Subsequently the CCL issued a third bulletin, entitled 'Support the Kirkland Lake Miners,' which advised that 'the strikers deserve unanimous and whole-hearted support from every Canadian worker.' This bulletin underlined the significance of the issue in the strike: 'The mineworkers of Kirkland Lake are fighting the battle for union recognition and for the very existence of the labour movement in Canada. If *any* employer can disregard the clear expression of their workers' desire to be represented by a particular union, and refuse even to meet a union committee and get away with it, no employer will feel that he ought to respect the right to organize and bargain collectively.' The CCL recommended that member unions bring the issue before their councils and local union meetings, and urged that provisions be collected for the miners and their families because 'the Congress regards this strike as one of great importance; the right of workers to organize and negotiate agreements through the union of their choice is at stake.'[23]

The congress had supported the strike cautiously and with a considerable degree of apprehension until the strike began, but thereafter the CCL increased its support. In November, the Toronto, Winnipeg, and Vancouver labour councils all passed resolutions condemning the government's labour policy and supporting the cause of the Kirkland Lake miners. Before the strike began, the *Canadian Unionist* had forecast that 'you will be hearing a great deal in the near future regarding Kirkland Lake; it is the focal point of labour activity in Canada at the present time.'[24] The events of the next few months demonstrated the accuracy of this prediction.

As the organizing campaign progressed and the CCL became more deeply involved, so did the CCF. The party was sympathetic to the objectives of organized labour and throughout 1940–41 was actively courting labour support,[25] especially in Ontario. On 18 November 1940, Clarie Gillis (CCF MP, Cape Breton) reported to the House of Commons that on his recent trip to Northern Ontario he had observed 40 000 miners who were 'endeavouring to form a union.' He attacked the government for failing to enforce PC 2685: 'What value is a statement of principle when in practice the machinery of the state and war regulations are used to thwart labour organization and to arrest trade union members?' Gillis believed that, in time of war, it was especially important to recognize the interests of employees because 'the right of workers to organize, the enforcement of social legislation, the protection of civil liberty, freedom of discussion are all things that are inherent in democracy. To talk about them is not good enough. They must be put to work as a practical demonstration to those in whom we expect to produce the sinews of war and to fight the war.'[26] Gillis was concerned not only with the absence of a positive labour policy, but also with actual intimidation of employees who sought to organize. On the day a strike vote was being taken at a Chrysler plant, soldiers marched through 'inspecting' the plant, and police were stationed outside the union hall throughout the vote. Trade unionists who sought to organize employees were arrested under the Defence of Canada Regulations. These events suggested not a neutral attitude, but rather an active policy designed to discourage union organization and collective bargaining.

Throughout 1941, CCF organizer Grant MacNeil travelled throughout Ontario developing trade union contacts, and becoming more and more deeply involved in trade union matters.[27] A conference of CCF and trade union officials held in early 1941 resolved to educate union members about the political objectives of the CCF, to establish a labour conference in Toronto 'for [the] purpose of consultation on legislative matters,'[28] and to impress on CCF members the desirability of joining trade unions. Feature articles on labour matters appeared frequently in the party newspaper, *The*

New Commonwealth. At the time of the strike vote, held persuant to PC 7307, the provincial CCF issued a statement condemning the new order in much the same terms as the CCL had done; and a month later the CCF protested the detention of Reid Robinson who had been imprisoned by immigration authorities while on his way to Kirkland Lake.

At a time when distrust of government was growing, the CCF and the labour movement were developing similar political objectives. The CCF's appraisal of the 'labour problem' contrasted sharply with that of the federal liberals, and the CCF was an early and vocal advocate for the introduction of Wagner Act principles into Canadian legislation. It took on the task of interpreting the new movement to the Canadian public. As a result, the CCF and the labour movement (particularly the CCL unions) began to forge a political alliance. From its position in the House of Commons, the CCF maintained a continuous attack on the government's inadequate labour policy and the application of that policy in particular industrial disputes.

In November 1940, Angus MacInnis had used the example of the Teck-Hughes dispute in order to highlight the inadequacy of PC 2685, which was 'just window dressing and accepted by the government as being just that.' He pointed to the lack of labour representation on war boards in contrast to the numerous business appointees in influential positions in the Department of Munitions and Supply. These men knew a great deal about industry, 'but they know nothing at all about democracy.' Organized labour resented their intervention and the anti-labour attitudes they inevitably brought into the government.[29] A year later, immediately prior to the (second) strike vote, MacInnis repeated his attack on the government's labour policy, which he considered 'altogether wrong,' and 'definitely hostile to organized labour.'[30] In support of this point he reminded the House of the National Steel Car dispute in Hamilton, in which the PC 2685 principles had been overlooked, even in an industry controlled by the government! C.D. Howe, the minister of munitions and supply, was accused of fostering company unionism for stating, at the end of that dispute, that any number of organizations could be recognized in an industry. Such policy would make it impossible for workers to organize and stabilize labour relations, for it implied government support for a number of fragmented and weak employee committees. This was the antithesis of the exclusive bargaining agency concept embodied in the Wagner Act.

The CCF position on union recognition was similar to that of the CCL: 'So far as union recognition is concerned all that should be necessary is that the union is a recognized union, that the majority of employees in a particular branch of industry are members of that union, and want it to act as their bargaining agent.'[31] Recognition was not a question for conciliation and arbitration; it should be an automatic right. MacInnis thus echoed the conciliation board report of the Kirkland Lake dispute.

These views were echoed by Dorise Nielson, who also addressed the House on the eve of the Kirkland Lake strike, and urged the government to follow the example of Britain and the USA and adopt policies that would ensure 'that private industry does not run roughshod over the democratic rights of the workers.'[32] She, too, emphasized that while the miners did not want to strike, they felt that their right to union recognition was not receiving the protection the government should be prepared to give it. There had to be equality of sacrifice during the war. If labour was expected to work for the war effort, employers should be required to recognize their employees' unions. No employer should be allowed to refuse recognition because it did not like the constitution of a union; PC 2685 had to be enforced.

Throughout the Kirkland Lake strike and afterwards, 'progressive' MPs hammered away at the inconsistency inherent in the government's policy of non-intervention and 'voluntary' collective bargaining. Employees were compelled to accept wage controls (in fact, PC 8253 had excluded salaries and executive compensation); employees were compelled to resort to conciliation; employees were compelled to accept government-imposed strike votes; and in the Arvida dispute, the government had responded with troops and hastily drafted cabinet orders. Yet when employers flouted the principles of PC 2685 and even walked out of the conciliation proceedings, nothing was done.

When the results of the strike vote were made public on 10 November 1941, Bob Carlin sent a letter[33] to all the mines urging them to open negotiations before 12 November and warning that the morning shift would not report on 13 November in the eight mines whose employees had voted to strike. On the same day, the local's president, William Simpson, informed McLarty that Local 240 had set a strike deadline for 7:00 PM on 12 November and wrote that 'in the interest of national and community welfare, we urge your department to use its good offices to effect a peaceful settlement of this dispute.'[34]

When faced with the prospect of an immediate strike, the Department of Labour intervened and asked the union to postpone the strike to allow a joint conference to be held in Kirkland Lake on Monday, 17 November. The union agreed to this suggestion, but proposed that the conference be held on Friday, 14 November, with the minister of labour in attendance. The minister insisted that the conference take place on the Monday, since Parliament was meeting on the Friday and it would be impossible for him to come to Kirkland Lake before that date. Apparently the mine operators had refused to meet in Kirkland Lake but had agreed to meet the minister in Ottawa. This posed a serious problem for the minister, since he had allowed the union to believe that he would meet the parties in Kirkland Lake and, on the strength of that undertaking, the union had postponed the strike. By the time McLarty replied to the union to insist on the 17 November date he had become aware of the operators' views and had decided to withdraw his earlier commitment. Accordingly, he adopted a policy of calculated ambiguity, which further damaged the credibility of both himself and his government.

McLarty advised the union that he would come to the proposed conference in Kirkland Lake on the Monday 'if circumstances permitted'; otherwise a representative of the department would be there in his place. He then mentioned that the 'operators had *agreed* to come to Ottawa' (not that they had insisted on Ottawa), and that he would have to get their agreement to any *change* in the program.[35] It was not, of course, the union that had sought a change, as McLarty's phrasing implied. The minister was accordingly vague about his ability to attend the conference in Kirkland Lake, but intimated that there would be no problem if the conference were held in Ottawa. Obviously the minister's intervention was the ingredient that might make such a meeting productive, so the union really had no option but to agree. McLarty had been able to convince the mine operators to meet with him and, the union believed, also with them. Such a face-to-face encounter might result in a settlement; the locale of the meeting was really unimportant.

Initially, however, the union only confirmed its agreement with the Monday date, omitting any mention of the place. 'The union had meetings all night ... as each shift came off,' and decided to defer strike action until a three-way conference could be held on Monday, 17 November in Kirkland Lake. At that time, the union indicated, it would be pleased to meet with McLarty or, if necessary, a department representative 'to work out [a] collective bargaining agreement on the basis of the conciliation board report.'[36] The union was reluctant to send its officials to Ottawa on the eve of the strike, and McLarty was therefore forced to communicate his actual intention of having the meetings in Ottawa. He told the union that such an arrangement was in the best interests of avoiding a strike. This would postpone the strike yet again, but it was extremely difficult for the union to reject the minister's request.

Carlin initially replied that it was impossible for the local union executive to leave Kirkland Lake 'in view of the acuteness of [the] situation here and of [the] imminence of strike action.' He protested the further delay. He assured McLarty that the union was anxious to co-operate 'in every reasonable respect,' but that the union did not regard the minister's suggestions as reasonable in the circumstances. The minister had had every opportunity to confer with the parties earlier. The union had systematically proceeded through all the proper steps prior to the strike and had resorted to all the dispute settlement mechanisms. The department itself had been involved for months.

The union earnestly wished to avoid a strike and reluctantly agreed to send key executive members[37] to Ottawa. Nevertheless, the union was bitter about having to concede to the government and to the mine operators. The meeting in Ottawa, if unsuccessful, would result in further delay, inconvenience, and expense.

Ultimately the union representatives went to Ottawa because they believed that the minister intended to bring the two parties together for the purpose of bargaining. The meeting represented the only possibility of settling the dispute. The union did not want to appear intransigent if there was a real chance of settlement, and it was angry when the press portrayed the union as being unco-operative. McLarty had proposed the meeting in Kirkland Lake but had changed his mind when the mine operators had advised him that they would meet only in Ottawa. He had indicated that because Parliament was sitting it would be inconvenient for him to leave Ottawa, but Parliament was not, in fact, in session.

The union's lawyer, J.L. Cohen, was infuriated by the conduct of the minister. He wired McLarty suggesting that it would have been a simple matter for the department to correct the press's erroneous view. He felt that the department's acquiescence in, if not actual encouragement of, these misconceptions prejudiced the conference, 'which if it is to be successful will have to deal fundamentally with fundamental issues.'[38] In Cohen's view, management had been able to make it appear that the union had flouted the government's wishes. McLarty had allowed the facts to be misrepresented and had permitted the companies to secure a public relations advantage.

The union was not optimistic. Before leaving for the conference the local union executive instructed the delegation that unless negotiations had begun by 3:00 PM on 18 November, they were to withdraw from the talks and return to Kirkland Lake.[39]

The conference was a great disappointment. The meetings never did discuss basic issues; in fact, it was not a 'conference' at all. Prior to the meeting, one union official expressed his expectation that 'at least the mine operators must now sit down in the same room with the union to discuss the problems of the miners – something they have hitherto refused to.'[40] He was wrong.

The 'crowning absurdity'[41] was that the parties met with the government separately: 'Even at this stage, the operators refused to meet in the same room as the miners' reps and the government could or would do nothing to bring them together.' Bob Carlin remembers that 'they were sitting in one room and we were in the other. I suppose if we wanted to listen we could hear them talking. They could with us. We were willing to go into their quarters, but they were not willing to come out to talk with us.'[42] And the government 'clothed in all the powers of the War Measures Act'[43] remained unwilling to act.

When McLarty met the mine operators' delegation, they presented him with a brief to explain their refusal to recognize the union. The brief contained nothing new, but was simply a repetition of all the old arguments: CIO irresponsibility, communist domination, improper organizing tactics, and a general assertion that alleged employee grievances were a fabrication of outside agitators.[44] The employers concluded their presentation by declaring that they did not oppose collective bargaining, but refused to recognize Local 240 because of an obligation to the companies' 73 000 shareholders.

The union in its meeting with the minister complained of dismissals for union activity and the hiring of strike-breakers, which, in its view, indicated that the employers had no intention of seeking a settlement. After listening to these grievances, McLarty proposed that both sides should seek a compromise. This first day of meetings had confirmed the impasse.

When the local union heard that its delegation was not even talking to management, it (correctly) assumed that no progress was being made and set 7:00 PM on 18 November as the new strike deadline. As tensions rose in Kirkland Lake, Councillor Anne Shipley urged the workers, over the local radio station, to accept employees' committees and stay at work.

On 18 November 1941, McLarty and the minister of mines, T.A. Crerar, met first with company representatives and then with the union. It was suggested by CCL President Aaron Mosher that agreement might be reached if he, Pat Conroy, and M.M. MacLean of the CCL met the representatives of the mines. McLarty stated that he would endeavour to arrange such a meeting, but by this time several of the seven management representatives had left for Kirkland Lake and only three were immediately available. These three agreed to meet with the minister, but 'it was proposed to contact [the] other four as well.' In the meantime McLarty was advised that the strike had been called and that the men were not returning to work. 'The mine operators felt that in view of this fact, no further negotiations would be of use.'[45]

McLarty told the three available management representatives that the union was willing to abandon its demand for a master agreement and was prepared to deal with each company individually. He also raised Mosher's suggestion of CCL involvement, which might induce further concessions. But as McLarty wrote to King: 'These proposals I immediately submitted to the employers. They advised me that it was not acceptable.'[46]

By the conclusion of these meetings, the mine operators' position was clear. They would agree to work with the employees' committees, which might perhaps include union members, but they refused to negotiate with the union: 'They said that they would meet with a committee of their own employees ... that's the position they took in Cobalt and for years it sort of worked for them. But they would not meet with any given union, and particularly the International Union of Mine, Mill and Smelter Workers, because it was a foreign-controlled organization, although practically all of their mines was controlled by foreign capital.'[47] This was their position after the IDIC investigation and it remained unchanged. If the union members proved to be 'responsible' and the terms of any agreement were observed, the companies would later *consider* agreeing to the participation of union officials in the discussions. There was no guarantee, nor was there any under-taking that employees who engaged in union activity would not simply be fired, as they had been in the past.

McLarty apparently believed that he was making progress, although neither of the parties to the dispute felt this way. Nevertheless, when the meetings broke down, a controversy erupted over who was at fault. McLarty told the press that efforts to reopen talks were continuing on 18 November and he was trying to arrange a smaller conference between a few mine managers and the CCL officials. However, while these manoeuvrings were continuing, management learned that the union had set a new strike deadline. As a result, management left Ottawa, blaming the union for the breakdown in the talks. What McLarty did *not* reveal was that most of the employer delegation had *already* left to prepare for the strike, and that those who had remained had rejected Mosher's suggestion immediately after it was presented.[48]

The CCL immediately replied to management's statement, which blamed the union for the breakdown in the talks, and to McLarty's version, which took the same view. The congress contended that, in fact, no progress was being made and that management's previous position had not changed. The CCL claimed that McLarty had told the union that the operators would nego-tiate only with employees' committees, and if after a year the individual committees were unsatisfactory, the employees could select a panel of employees, *who were agreeable to management*, to represent them. If this assertion was correct (and the mine operators denied saying it), it repre-sented a retreat from even the minimum recommendations of the IDIC.

The congress claimed that it was the minister's duty to publicly declare his disapproval of the mine operators' 'undemocratic behaviour' in flouting the government's labour policy. The congress claimed that he had privately expressed this view, but the Department of Labour denied that McLarty had said the operators were not in compliance with the government policy.[49] The union itself blamed the breakdown on management intransigence; it observed that when the talks broke down, management representatives were already leaving Ottawa and the union was unaware of any further proposed meetings.

What is clear from this series of recriminations is that the meetings accomplished nothing but a delay of the strike. The government's mediating role did not induce the parties to discuss their differences. The government was still trying to work with the IDIC proposal and was unwilling to apply any pressure that might break the stalemate. Ultimately neither party would make concessions. The structure of bargaining was *the* basic issue and there was little room for compromise, particularly in view of the distrust that had developed in the period prior to the strike. The government was still unwilling to consider any basic change in its labour policy. The Kirkland Lake Formula was the maximum the employers would tolerate, and was the solution preferred by the government. Since it meant the abandonment of their union, it remained entirely unacceptable to the employees.

As the strike began, the Canadian Congress of Labour issued a statement pledging 'every assistance' to the Kirkland Lake miners who 'faced the alternative of submitting to the dictatorial attitudes of the operators or peacefully withdrawing their labour.'[50] In the absence of a Canadian Wagner Act or federal government intervention, the employees had two options: to give in to the mine operators or to strike. They struck.

7
The strike

On the evening of 18 November 1941, the night shift in eight of the mines failed to report for work. Peaceful picketing began at the mine gates. At 6:00 AM the next morning more pickets arrived; and 'many women were on the picket lines with their men.'[1] Kay Carlin, president of the union's ladies' auxiliary, recalls that 'the night it broke, we [the women] were all on Government Road in front of the Lake Shore mine in a pouring rainstorm. Everyone of us were out there ... And it was just *pouring* rain the night they went out ...'[2] The *Star* reported that 3850 of 4300 miners had struck.[3] The union issued a statement that 'no production workers on the day shift reported for work and all the mines are at a standstill.' The operators countered this union report (as they would do all future union statements) with their own assessment, claiming that the Macassa and Toburn mines were operating as usual, and that the Sylvanite, Teck-Hughes, and Wright Hargreaves mines were running on a reduced scale. It was admitted that the Lake Shore and Kirkland Lake Gold mines were shut down completely.

As soon as the strike began the strikers and their families began to organize for what they correctly anticipated would be a long and bitter struggle. Their situation was similar to that of other strikers, except that this strike had to be fought through the ferocity of a long northern winter.

The local union officers and their committees were in charge of the administration of the strike. The union distributed food vouchers, in accordance with an arrangement previously negotiated with the local merchants (as was the later extension of credit for those men rehired at the end of the strike). The large stores generally refused to accept these vouchers, but the small stores agreed to do so. The vouchers were valued at about $8 per week, and enabled the striking men to buy essential food items for their families. The union believed that this system would ensure that the money was spent on food rather than on liquor. The union tried unsuccessfully to persuade the provincial government to close the local liquor stores for the duration of the strike. The union also tried to persuade the local taverns to close periodically during the day and were supported by the municipal council in this matter. The local was acutely aware that prolonged idleness might lead to drinking. Excessive use of alcohol might provoke disorderly conduct, which would reflect on the character of the strikers and the reputation of their union: '[we have] pursued an absolutely peaceful policy all through the strike. We have asked and repeat the urgency of it now, that all beverage rooms, liquor stores, wine stores and brewers' warehouses should be closed up immediately. There already have been a couple of incidents caused by men being under the influence of alcohol.'[4]

As the law then stood, strikers were entitled to picket peacefully in order to communicate information concerning the strike. They were also entitled to address the workers crossing the lines and attempt to persuade them not to go to work. Picketers were *not* permitted to use threats, or violence or intimidation, but as in most volatile strike situations, the line between 'persuasion' and 'intimidation' is difficult to draw. Name calling, insults, and other forms of verbal harassment are inevitable. So long as there is no violence, however, this conduct is not illegal, and the expressed enmity of their fellow workers is something that strike-breakers must learn to endure.

The union had to administer these picketing duties, organize the thousands of strikers into shifts, and allocate the pickets to the different mines. Picketing was especially important where management planned to continue operations. It provided clear evidence that the miners were united in their support for the strike and would act as a deterrent to any potential strike-breaker. In the tight-knit northern mining communities this social pressure was an effective deterrent so long as the strikers' solidarity could be maintained. In addition, picketing provided something for the idle workers to do, and maintained their contact with the union so that new developments could be communicated and group cohesion maintained. The success of the strike ultimately depended on effective picket lines.

From the company point of view, it was essential to frustrate the unions' picketing campaign, for it could be very effective in the context of a company town where employees would not wish to be subjected to social pressures from their fellow workers. For these employees the stigma of 'scab' or 'blackleg' was a serious matter. Allegations of violence, however, would undermine the legitimacy of the union's conduct, and might even provoke indignation and actual violent incidents. This would disrupt the picketing and disorganize an important union device for maintaining contact with the striking workers and conveying information and encouragement.

The union denied management charges, and portrayed picketing as orderly and peaceful as, initially, did most of the press. The union also denied that men were returning to work. The union alleged that 'the companies are most anxious to create an atmosphere that will embarrass the local authorities in order to give a certain union-hating provincial leader an opportunity to interject himself into the dispute.'[5] Eventually the union had to establish relations with the provincial police (OPP) sent in by the Ontario government *one week* after the strike began. The police became the union's adversaries on the picket lines. After the OPP arrived, there were strict rules about picketing, and initially the police banned all picketing of the rear gates of the mines. After the union complained that strike-breakers were entering by these gates, it was permitted to resume limited picketing. In the circumstances, some picket line 'incidents' were inevitable. The union had to keep track of the charges against its members, appear with them in court, assist their counsel (J.L. Cohen) to gather evidence for their defence, and sometimes raise money to pay fines.

The strikers themselves had definite duties. Each man was required to do several hours of picket duty a day. Tents were erected by the mine gates to provide shelter. These were equipped with stoves and a supply of fuel.[6] When the men were not picketing they were assigned to woodchopping and the collection of firewood in the forests. The women helped to pile this wood, which was then supplied to families according to their need for use in their wood stoves.

The strikers and their families paraded through the town at important junctures during the strike to protest various issues (such as the number of OPP sent in by the provincial government), to demonstrate solidarity, or to combat management and newspaper efforts to start a back-to-work movement. Such parades involved between one and three thousand people.

The local organized its own fund-raising projects to supplement the assistance it was receiving from other local unions across the country. It was decided that their young recording secretary, Larry Sefton, would travel across the country to speak at local union meetings, to 'tell the story' of the Kirkland Lake strike and to solicit financial contributions to the strike fund. As Bob Carlin said: 'We both [were] termed about equal as far as experience is concerned and qualified to do that work, to address meetings. Neither of us were public speakers but we ... had a fairly high degree of sincerity and certainly were wrapped up in the Kirkland Lake strike. If we couldn't talk about anything else, we could talk about the Kirkland Lake strike. So Larry was chosen to go out.'[7] It was hoped that a personal appeal by a young miner would increase the already generous contributions that were beginning to be received from locals who supported the strike.

Sefton was gone for about a month, travelling from Nova Scotia to British Columbia. Everywhere he went the pattern was the same: the locals billeted him, he spoke to their meetings, the money would be contributed, he would take enough out of it for his fare to the next town with a local union in it, and would send the rest back to Kirkland Lake.

Local 240 had an active 'ladies' auxiliary.' These ancillary organizations were established by many locals in the realization that a union could never be organized if there was opposition from the women.[8] During the union's organizing drive, women in the ladies' auxiliary did a house-to-house canvass to persuade the wives and mothers of union members to join the auxiliary, and to inform the women of the union's objectives. The auxiliary proclaimed that through the organization of Local 240 'our men are now able to win for us the extra wages and the safer mining conditions that have been needed in Kirkland Lake for many years.' When the companies appealed to the women to urge their husbands to join the workers' councils, the ladies' auxiliary wrote to explain that 'these workers' councils are paid for by the companies and they think a good way to get at our men is to make a cheap and insulting appeal to us women to desert our men.'[9] Prior to the strike the auxiliary explained that the mine operators had persistently refused to negotiate and assured the women that in the event of a strike 'the union will see that no union member or his family suffers,' and would provide strike benefits.

The auxiliary (which met once a week in the union hall) urged all of the women to become members: 'It's the very best way we women can show that we stand by our men.'[10] The women of the ladies' auxiliary helped maintain the union hall and undertook 'to do the best we could in assisting the miners' union ... [in] whatever program and plans they had.'[11] The auxiliary did not participate in the actual collective-bargaining decisions taken by the membership of the local.

The auxiliary became the vehicle for providing food and coffee to the picketers: '[At] every mine, we all had our shift. We had four hours on and four hours off and we would rotate ... we would make the coffee and we would bring it out.' It was sometimes forty below zero on the picket line, 'so we would go out on our shifts and bring them coffee. Even if we had to walk, we went out there ... Our job was to service the union and ... we went out to help the guys on the picket line.'[12] To free some of the women for these tasks the other women looked after the children, ensuring that in this situation of scarcity, each child received enough food and milk and a daily portion of cod liver oil.

The women had other jobs as well. As bales of food and clothing arrived from around the country, the women helped the local's Strike Relief Committee sort and distribute the gifts. Mrs Carlin recalls that the neighbouring farm communities around Cobalt and Haileybury sent beans to the strikers with a note that read, 'that's all we have.' These were received gratefully and were divided among the families. She also recalls that 'when Christmas rolled around, the Timmins local sent down a *ton* of Christmas cake. One ton of Christmas cake and that was divided up between all the families.' The women even provided entertainment for the striking miners and their families throughout the strike, and helped to pass a rather bleak Christmas: 'They came in and they played old Irish songs ... And the children even used to come out and have choirs and they would sing in the Union Hall, put on little dances and things like that ... And we would all gather there at night and see what we could do ... to cheer up people and keep them going.'[13] These impromptu concerts and the organized entertainment provided a source of relaxation in a difficult situation. As has already been noted, it was crucial for the success of the strike that the strikers remain active, since idleness would have a particularly demoralizing effect among these men who were so accustomed to hard work.

Throughout the strike the ladies' auxiliary continued to inform the women about the union: 'We knew its program, we knew its policy. We listened to their speakers. If anybody came, we rounded up people to get out and listen to them.'[14] One long-term effect of such activity was its 'politicizing' effect on the miners and their wives. The auxiliary actively worked in the federal and provincial election campaigns of 1943 and 1945, and contributed significantly to the election of CCF members.

POLARIZATION OF THE COMMUNITY

The community of Kirkland Lake was shaken and divided by events during the strike. There was widespread fear that the strike would ruin the town. The *Northern News* expressed a common 'middle-class' view when it warned that a strike would cause some of the mines to close indefinitely, that business in the town would be hurt, and that 'nothing but harm, loss and the disruption of the community can follow ...' The publishers expressed the hope that the businessmen in the area would 'give guidance to the more sober men of the miners' ranks. Many of them are praying for just such assistance.'[15] The *Northern News* accepted the naive theory that CIO professionals and outside agitators had manufactured grievances and used these to deceive the miners into supporting the union. As Dr Mutchmor pointed out to an anti-CIO minister in Blenheim, Ontario, who was concerned about the Kirkland Lake situation: 'Do you think that 4,000 men are going to give up good jobs at the beginning of the winter season, supply pickets for [a] 24 hour continuous outing in zero and sub-zero weather and in other ways endure hardships just because some smart chap comes and tells them to go on strike? Our ministers resident in the Kirkland Lake area do not think that the miners went out on strike because of foreign agitation.'[16]

Mr P. Bennett, the electrical superintendent of the Wright-Hargreaves mine (who was opposed to the strike), considered his own view an 'independent common-sense opinion.' He spoke to the townspeople over station CJKL just before the strike began, advising that the strike would bring economic hardship to families, some mines would be shut down permanently, and there would be a loss of jobs and a loss of revenue for the town.[17] There was a real concern that relations among the townspeople would deteriorate permanently. Bennett in his radio address pointed to 'a gradual loss of fellowship' in the last two years.

The Reverend E. Long, the local United Church minister, and a less biased observer, also commented periodically on the mood of the town in correspondence with his colleagues. After the conciliation board report was adopted by the union and the first strike deadline was called, he noted that rumours were flying, the atmosphere was getting tense, and there was the feeling that a strike was inevitable within the next few days. There was the 'feeling we are sitting on a keg of powder here.'[18] The government-supervised strike vote took place in an atmosphere of surface calm, but as the final meeting with the minister of labour was being arranged, he wrote of 'a very sober town now – everyone is anxious.' Once again rumours abounded. One such rumour – much to Reverend Long's distress – was that the United Church ministers were crusading for the union at union meetings.

There was an obvious desire to avoid ill feeling among people who had to live in close association. However, social stratification was an important factor in the Kirkland Lake dispute. It defined attitudes and loyalties during the strike, and contributed both to the division of the community over the strike and to the solidarity of the miners. Those who feared the strike persisted in seeing it as the result of the 'manipulation of a minority of agitators from across the line who care nothing for the tradition of good Canadian hard-rock mining,' – 'whose sole purpose was to stir up trouble in one of the highest paid, happiest camps in Canada.'[19] These community spokesmen sought to maintain the *status quo* or to introduce employees' committees.

The unionized miners' solution to the deterioration of labour relations in the community was the introduction of collective bargaining with an independent miners' union. This institutional change was seen by the mine operators and their sympathizers as a bid to 'take over the mining industry,' and while their view was exaggerated, there is no doubt that collective bargaining would have fundamentally altered the power of management to make decisions without regard for their impact on employees. Positions were polarized and 'the acid of the bitterness had eaten its way into the heart of the community.'[20] Families quarrelled as some members went on strike while others kept working. For years afterwards, men were labelled by their co-workers according to the position they had taken during the strike.

The professional classes in Kirkland Lake were generally sympathetic to the mine operators. This was not surprising. The first professionals had been attracted to Kirkland Lake by the mining companies to develop the mines or (as in the case of doctors) to service the growing mining communities. There remained a close relationship between such people and the companies. Lawyers were opposed to the strike because they would lose business, and were, in any event, disinclined to favour collective action. Teachers were anxious about the probable loss of revenue to the town and the disruption to the educational system. All of these people shared the management perception of the CIO that was continually pressed in the local media.

The business community was divided. The financial institutions and large chain stores were sympathetic to the employers. A substantial number of the smaller merchants, whose customers were the striking miners, supported the strike. Before the strike began, McLarty received a resolution signed by a meeting of key businessmen in Kirkland Lake asking the federal government to insist that the mine operators negotiate with the union.[21] During the strike, seventeen small merchants – grocers, dairymen, and bakers – extended the usual two weeks' credit to their usual customers who were then on strike. Many of these businessmen also accepted union vouchers for supplies and extended credit to the miners after the strike. Those who were taken back were given an additional two weeks of credit until they were paid. The extent of the contribution of these small local merchants was acknowledged by Local 240 after the strike. The local requested funds from other locals to pay their debts to the small merchants 'whose warm sympathy and ready credit carried the strike along when financing became difficult.'[22]

The managers of collection agencies in Kirkland Lake were predominantly opposed to the miners' struggle, since they too had close connections with the mining companies. However, at least one spokesman publicly admitted that the recent boom in their business had occurred because the miners' inadequate wages made it necessary to go into debt.[23]

Those who opposed the miners and their union recognized that there was more at stake than a simple demand for better terms and conditions of employment. One underlying theme of the local struggle (and indeed of the national struggle of the labour movement as a whole at this time) was the miners' desire to gain 'recognition' – individually as persons, collectively through their union as a partner in industry, and nationally as part of the labour movement and a partner in the war effort. 'The miners insisted upon being treated as mature and responsible citizens in industry and in society.'[24] Labour spokesmen declared that workers could no longer be viewed as a commodity in the industrial process and rejected the view that employees were being manipulated by 'outside agitators.'

To some extent their strike was a struggle for a new status in the local community. The mine operators, who regarded the mining communities as 'their' towns, which they had built and paid for with 'their taxes,' now faced organized employees who valued their own contribution as productive citizens and more recently as war workers. They too had contributed to the community, paid taxes, and served on local government bodies. The union became the vehicle to acquire a new status.

The middle class[25] of the town rejected this change, in part because of the changing ethnic composition of the community. Unwillingness to accept newcomers reinforced middle-class reluctance to recognize or support a change in the industrial or social *status quo*.

The standard bearer of the anti-union, anti-strike sentiments in the community was Anne Shipley – a city councillor and candidate for reeve at the time of the strike. The split in opinion in the community was reflected in municipal politics and in the make-up of the Teck Township Council. Shipley spoke for the middle-class anti-strike adherents, while Reeve Carter tried to conciliate the two sides. When under pressure, however, he proved to be sympathetic to the miners.

When a municipal election was held on 1 December 1941, the work stoppage became the major issue. 'The strike issues have overwhelmed all else in the election and the votes will be polled undoubtedly on a pro- and anti-union basis,' reported the *Globe and Mail*. In the election campaign, Reeve R.J. Carter was seen as the strikers' candidate. Mrs Shipley and two other candidates ran on an 'anti-CIO' platform.

The results of the election demonstrated the division in the community. Reeve Carter was re-elected and won narrowly over Mrs Shipley by 3440 to 3122 votes. The council members changed and the council's allegiances went from a four-to-one to a three-to-two pro-union majority. Councillors H.J. Symington and A.L. Trudeau opposed the strike and the union. C.C.

'Doc' Ames and Tom Church favoured the strike and were active union members.[27] The union found the election results 'gratifying' and interpreted Mrs Shipley's loss as public endorsation of Local 240.[28] This was an exaggeration. What the election had demonstrated was a high degree of polarization within the community.

The churches also felt the repercussions of the conflict, for as in most small towns, they were an important force in the life of the community. As Lucas observed in his study of one-industry towns:

The church is the focal point of much social activity in a community of single industry; under its umbrella many voluntary associations serve various age groups ... a great number of the activities for the town's young people, much of the choir music, women's organizations, teas and suppers are carried on under religious auspices. The churches provide a ready made in-group, as well as the all-important physical facilities for club activities. In this way, then, the members of a denomination are brought together in active interaction; cliques of friends and close associates form along denominational lines.

Lucas found that denominational differences divided the community, especially when they were associated with ethnic differences; however, he also noted that 'the churches provide a regrouping which is different from the structuring of the company hierarchy. They also provide the bases for a social life and cliques that go beyond either association within occupational level, neighbours, or community-sponsored recreation.'[29] Thus, *within* each denomination, the churches provide a basis for common belief and social interaction that may transcend occupational or class differences.

In Kirkland Lake the church organizations were the one place that employer and employee representatives had both held prominent positions. In contrast, property and economic relations in the town reinforced class distinctions and exacerbated conflict, since in a one-industry town consumer, landlord/tenant, and employer/employee grievances were all directed against a single source of frustration.

In normal times the role of the churches was to minimize the divisions in the community based on occupational differences: 'Although the churches provide in-groups and out-groups, social life is so structured that there is seldom open conflict.'[30] With the advent of the strike, the possibility of open conflict was very real. The various congregations struggled against it; since once again another pattern of traditional relationships would be disrupted. In order to avoid disruption in their congregations they stressed the 'Christian gospel' of co-operation and tolerance. Many of the churches felt that this objective could best be accomplished by supporting the mine operators and urging the workers to join the employees' committees. One church split over the controversy. Another, the United Church (which was the largest Protestant church in the area), experienced prolonged internal struggle but was successful in maintaining the integrity of its congregation by 'agreeing to disagree,' and adopting a generally neutral position.[31] Ultimately, most of the churches had to take one position or the other, or were perceived as doing so by the community.

The Roman Catholic church split. The Irish Catholic priest, Father Cavanagh, was strongly sympathetic to the strikers. After one of the parades through the town, he phoned the local organizer, Bob Carlin, and remarked: 'Metaphorically speaking, you gave them a great kick on the seat of the pants when you took that parade. God bless you, you and your strike.'[32] In contrast, the French priest acted as a 'scab-herder' on behalf of the companies. The union discovered this when a French-Canadian strike-breaker was brought to the union by a Slavic striker who had been picketing the back part of one of the mining properties. Because local workers were unwilling to 'scab' on their fellow employees, the companies had been recruiting in other mining towns in Ontario and Quebec. This particular strike-breaker was carrying a note from the priest addressed to the company, testifying to his good character. The priest had apparently talked to him and assured the company that he would be 'loyal.'[33]

As a result of this incident, a three-man delegation visited the priest. At first he denied that he was acting as a 'scab-herder,' although he expressed extreme antipathy to the strike. Finally, when presented with the evidence of several such notes, he admitted that he had been recruiting strike-breakers. The union agreed to say nothing if he immediately ceased this activity, but threatened to publicize his conduct if it were not discontinued. There was no further trouble from that quarter.

What was particularly galling to the union was that those men who had been recruited by the priest, and who had knowingly walked into a strike situation, were doing so only because they were poor and had been unemployed for years. They had no particular antipathy to the strikers, but were driven by economic circumstances to take any job available. The priest's recruiting 'success' depended on the poverty and desperation of the unemployed miners from other towns. From the union's point of view it was a form of exploitation for all concerned. In later years, many of these strike-breakers became members of Local 240.

The Protestant churches took varying positions.[34] Three mine managers were Anglicans. This influenced Reverend Judd's refusal to become involved with the United Church ministers who supported the principles of union recognition and collective bargaining. The local Presbyterian Church split on the question. Their leading layman was the manager of the Teck-Hughes mine, but the president of Local 240 was also an active Presbyterian.[35] Once the strike was in progress the Anglican, Presbyterian, and Baptist ministers all publicly took the side of the mine operators. At least two spoke from the pulpits and favoured 'company unions' over the CIO.[36]

Of all the churches, the Methodist Church was the most pro-labour in its policies, and the most actively interested in industrial relations matters. In September 1941, prior to the conciliation board hearings, the Temiskaming Presbytery of the United Church of Canada published a report on the Kirkland Lake labour-relations situation. The seven-man commission included United Church ministers in the northern mining areas of Kirkland Lake, Cobalt, and Swastika, and was established because of the Temiskaming Presbytery's belief that 'the growing tension in the relations of labour and capital in Kirkland Lake has [the] potential capacity for disaster to community and individual life,' and that 'the welfare of the general public in the locality affected is of equal, and in some ways, of more importance than the matters in the dispute.'[37]

The commission investigated the local union and determined that the officers were all local people with the exception of Tom McGuire, the district administrator, who was an employee of the union. Of the one-dollar-per-month membership dues, sixty-five cents went to the *Canadian* headquarters of the union for 'promotion and organization.' The rest remained with the local union. No money left the country. The local's books were audited regularly and an annual statement of finance was sent from the Canadian head office to the local. The union demands had been drafted by five men chosen democratically from each of the ten operating mines to form a committee of fifty men. The report reviewed the steps taken by the union to get a conciliation board, which had not as yet arrived in Kirkland Lake at the time of the church investigation. The report concluded that 'the chief point at issue is the question of recognition of the union as the sole bargaining agency of the miners.'

The report was consistent with the policy of the United Church, which favoured the 'right to workmen to bargain collectively through representatives of their own choosing.' The churchmen believed that these rights were 'fortified by law' (i.e. section 502A of the Criminal Code and PC 2685), and they referred to statements by the minister of labour (Norman McLarty) that had stressed the necessity of employers to recognize the right of workers to bargain collectively through an agency of their own choice. The commission asserted that this right was recognized in practice in Great Britain and was supported in law by the United States Supreme Court. To neglect these rights in Canada would be 'prejudicial to the maintenance of the law generally among our people.' On the basis of these 'facts,' the report concluded that:

1 improved employer/employee relations required mutual respect between the parties to the dispute;
2 the role of the church was to maintain the basic human rights of people in the community and to act as peacemaker by pointing out to both sides the necessity of a conciliatory, co-operative attitude;
3 the United Church statement adopted in 1940 called 'Industrial Organization and Collective Bargaining' should be endorsed;
4 it was necessary for both parties to clarify their positions and particularly for the government to clearly outline the meaning of its war labour policy;
5 the delay in initiating the proceedings of conciliation boards might weaken labour's position;
6 neither party should take advantage of the war situation to advance its own interests. All individuals in Kirkland Lake should bring to bear on the immediate issues an 'enlightened Christian conscience, that the ultimate welfare of all concerned may result.'

The report was a conciliatory document, but it was ahead of the government in its acceptance of economic and social changes in society and the necessity of legal protection for employee rights. It was not as explicit as the United Church policy on labour relations. It neither enthusiastically embraced the 'new unionism' nor criticized management. While the local union and the directors of the United Church's Board of Evangelism and Social Service interpreted the report as favouring the recognition of Local 240, the members of the commission were more vague. Reverend Long of Kirkland Lake claimed, 'We upheld the right of collective bargaining, not necessarily Local 240, save as the general principle is involved.'[38]

Nevertheless, the report was greeted favourably by the union. The commission's investigation of the local's officers, finances, and decision-making processes demonstrated that the union was run responsibly and democratically. This image contrasted sharply with management's portrayal of manipulative foreign labour bosses who were running the local union for 'profit' in order to take over the industry. The church commission recognized the right of collective bargaining through a bargaining agency of the workers' own choice, and in addition interpreted PC 2685 as a government endorsement of collective bargaining. This was an interpretation shared by the labour movement, although it was becoming apparent that the federal government did not intend to enforce its own policy. The local commission was an independent body that was primarily concerned with the public interest. Its favourable report had considerable impact.

When the Temiskaming Presbytery's report was made public, J.R. Mutchmor, a secretary of the Board of Evangelism and Social Service, wrote Reverend R.J. Scott of Cobalt to commend the work of the commission. He expressed his own view of the situation: 'It seems to me that the government of Canada could very easily settle a lot of these labour problems by simply making the findings of these [conciliation] boards mandatory. If the boards had the power to order industrial concerns to recognize the right of men to collective bargaining through agencies of their own choosing, some of the problems would be solved.'[39] Both the local ministers and the church administrators in Toronto consistently supported this solution; but the central church office worked for a resolution of the dispute on the national level. The local ministers in Kirkland Lake and surrounding communities were primarily concerned that their congregations should not become totally divided over the dispute. To this extent, reverends E. Long, J. Boyd, and W. Cullis tried to remain neutral and reconcile the two sides. They believed that the miners had a right to a union, but they resented it when the union interpreted their position as unquestioning support for the local. They feared that such an interpretation would be divisive. When the strike was underway, neutrality was difficult to maintain – as J.R. Mutchmor recognized. He praised the local ministers for their efforts, recognizing their motivation was the preservation of their congregations. He and his assistant, Dr J. Coburn, however, were not neutral. They had 'strong views on the Kirkland Lake situation'[40] and believed that in time there would be no neutrals because 'the bitterness and hardship increases as the strike progresses.'[41]

Mutchmor and his Toronto staff were primarily concerned with the basic issue of collective bargaining and they continued to view the strike in a national political context. Mutchmor wrote: 'It is my firm, deliberate judgment that little or nothing can be done in Kirkland Lake. The issue now depends on what action the Federal government will take.' His assistant, Dr Coburn, held the same view. He wrote to Reverend Cullis: 'Remember this is not entirely a local situation. There is a great principle involved here in which the whole labour movement, and in fact all who desire a better social order are deeply interested.'[42]

The local ministers were embroiled in the personal and complex repercussions the strike was having on the community. Mutchmor's pro-union view, which he published in a letter to be read out to the congregations, seemed too partisan and contrary to the aim of his colleagues to keep their congregations in a state of 'undivided fellowship.' The local ministers' suggested that Mutchmor abandon his plans to visit Kirkland Lake.

After the strike began, most of the other churches in Kirkland Lake publicly declared their support for the mining companies' position. Reverend Boyd of Swastika reported to Mutchmor that this had caused the local ministers considerable concern. They were under pressure to abandon their 'sympathetic neutral' stance, and actively oppose the CIO. At this critical time there was considerable tension between the central office of the United Church and the local ministers.

The concern about the events in Kirkland Lake was not confined only to the local ministers and the central office in Toronto. On 9 December 1941, the Toronto West Presbytery passed the following resolution:

Resolved that this Presbytery ... urge the Federal government forthwith to take such action in the Kirkland Lake strike as will make good to Canadian labour the government's promise, that the right of collective bargaining through organizations of labour's own choice be effective to the end, that in the dispute in question much needed gold production be carried on without hindrance, and that in general Canadian labour be encouraged to an 'all out' effort so sorely needed in this day of ever-increasing danger and crisis.[43]

Reverend Coburn was not present at the meeting but stated, 'if I had been, I would assuredly have voted for it.'[44] By this time he supported unequivocally the principle of collective bargaining and believed that the mining interests and their political allies were trying to frustrate the efforts to organize. Despite the hostility of the press and large sections of the public, he was convinced that the church should take a position supporting the union.

The Toronto West Presbytery had adopted the resolution because of its own experience and knowledge. The area within its bounds was one of the most industrialized in Canada. In the weeks before the resolution was passed, three strikes had taken place – at the Campbell Soup plant, General Electric, and Acme Screw and Gear. 'This experience of industrial strife was much in the minds of the one hundred or more Presbyters, when they voted. Six persons opposed the resolution.'[45] The resolution was much publicized, but the ministers on the scene in Kirkland Lake reported that the views expressed in the resolution and in Reverend Mutchmor's circular letter had upset many local people: 'Many of our leading and loyal church members are mighty critical and dissension is noted in the fellowship of the church.'[46]

As the strike continued and some violence occurred, the local ministers (and particularly Boyd and Long) began to be more sympathetic to the employee-committee solution, since the goal of independent unionism seemed both divisive and unattainable. When this shift in opinion became apparent, Mutchmor and Coburn wrote to support the union and the principles for which the strike was being fought. Both men were sympathetic to the aspirations of organized labour and troubled by the criticism of the church emanating from some of its traditional supporters. In their view the church had a duty to support progressive social causes.

In Kirkland Lake, the three local ministers emphasized the 'right of individual conscience,' 'the freedom of the pulpit,' and the right of the church to lead in social matters; but none of them openly or publicly stated their convictions about the strike situation as their confreres in the other denominations had done. Ultimately, Reverend Long allowed his congregation to discuss the local labour situation in an effort to avoid an open schism. His conciliatory approach was successful. The debate concluded with a general agreement with the three principles above, and the discussion markedly reduced the tensions within his congregation. The congregation recognized that there were faults on both sides, and that co-religionists were sincere in their differences. All agreed that the government needed to appoint an impartial tribunal to conclude a just and lasting compromise.

The crisis in the local church passed. It became apparent that the dispute could only be settled by government intervention. Thereafter, the focus of the United Church's activity (in conjunction with that of other Protestant denominations) shifted to the national level. Nevertheless, the crisis in the congregations of the local churches indicated the impact that the strike had on community life.

With the exception of the *Toronto Star*, the media were generally opposed to the strike. The news presented in the *Globe and Mail* (with which the mining industry had close connections) was sufficiently slanted that the *Canadian Forum*, at that time a moderate left-wing journal, wrote an article entitled 'A Newspaper Fights a Miners' Strike.'[47] It accused the *Globe and Mail* of sensational coverage of the strike, intentional omissions from quoted remarks of the parties involved, incomplete and biased versions of the Canadian Press dispatches, and even fabricated charges that the strikers were committing violent crimes. The local press was equally hostile to the IUMMSW (mine mill) and to the CIO. The *Northern News*, the *Northern Miner*, and the press in Timmins and Sudbury were all pro-management. The *Northern News'* publisher, Richard Pearce, ran editorials 'in our campaign against CIO activities.'[48] Its 'news' accounts reflected the fear that the strike would prejudice the town's future. The *Northern Miner* was primarily a trade paper and its sympathy for the employers reflected its preoccupation with the profits of the shareholders.

The *Northern News* was vicious in its attacks on the union, and continually urged the miners not to strike. As a retaliatory measure, the union urged its members to boycott the paper, which caused it to lose business. The paper accepted the management view that the miners were being manipulated by outside agitators. It advised the men to ignore such 'CIO machinations' because they were well off and a strike would only do irreparable damage to Kirkland Lake. Throughout the strike, the paper urged the miners to go back to work.[49]

On 23 October 1941, after a series of eight hostile items between 19 August and 3 October 1941, the international union decided to begin libel proceedings against the *Northern News*. Two of the libelous editorials were titled 'Where Do You Stand?' and 'Are Such Men Fit to Speak for Kirkland Lake?' The former contained the following passage:

There is every evidence that the union heads are in the fight to the last miner of Kirkland Lake, to the last business, to the last home. They would chance disrupting a community of steady working, highly paid men, a prosperous community, to gain their own ends. They've nothing to lose – keep that big fact in mind. They'll loot this country just as surely as the Nazis are looting Europe – and their principles are not much higher for all their grand talk about the glorious rights of the workingmen. It's their own CIO rights they're thinking about – a dollar a month from every stooge in the country. We can imagine them in the back rooms – gloating and rubbing their hands with glee. What a sweet racket! No wonder they're sore at the *Northern News* for showing up their lies, and their schemes and their disloyalties to Canada and all good patriotic men and women.[50]

These bitter diatribes, caricatures in fact, continued throughout the strike. The union gave up any hope of having its side of the dispute presented in the established press.

In April 1942, Reid Robinson decided to drop the libel suit. This decision was taken against the advice of J.L. Cohen, who felt that it might be successful and that there would be bad publicity if the case were dropped. Robinson was concerned because the defence wanted to bring out his left-wing political connections.[51] He argued that the continuation of the case could create ill feeling and might disturb the unity of the war effort and the international co-operation between Canada and the United States. By that date he had adopted the Communist Party position, advocating allied unity for the prosecution of the war effort.

The union responded to the adverse coverage in a novel way. Realizing that it would never be fairly treated in the existing press, it established a competing newspaper called *The Northern Citizen*: 'We thought if we get a good paper in here, it could build itself up and when the strike is over, there'll be enough miners here and business people that got acquainted with this little paper that it would ... become a useful organ in the community. But it didn't. It died with the strike.'[52] A man named John Smaller, who had come to Kirkland Lake at the beginning of the strike, agreed to establish an 'alternative press.' He was a union man himself (belonging to the Newspaper Guild), had been with the Canadian Press news agency, and had set up his own press in the North. The union agreed to help him produce the *Northern Citizen*. One of the miners (and later trade unionist), Bill Sefton, helped to organize the paper. Throughout the strike this newspaper presented the union's position, and the opinions of those people in the community whose views were not represented by the established press.

Another indication that the union was aware of the need for 'good public relations' was the selection of Eamon Park as a full-time publicity officer. Park had been 'loaned' to Mine Mill by SWOC (Steelworkers' Organizing Committee). He drafted press releases and leaflets for the local, and prepared a daily local bulletin called the 'Bullet.' The 'Bullet' informed the members on the progress of the strike, kept up morale, and countered company communiqués concerning the number of men at work in the mines.

One particular issue of the bulletin gained national attention.[53] It reported that the Toburn mines parent company – the American Mining and Smelting Company – had concluded a new agreement with the Mine Mill union in the United States. This was their fifteenth agreement with that union. In reply to management charges that the union was 'foreign dominated,' the leaflet informed its readers that three of the eight mines had executive offices in the United States, five of the eight mines had a president or vice-president who was a resident in the United States, and that seven of the eight mines had American directors.

Perhaps the most vital communications instrument in the propaganda war was the local radio station, CJKL. Both the mine operators and the union broadcast on a daily basis. Periodically Mrs Shipley would assail the CIO. There were regular time slots alloted to each party. The local's recording secretary, Larry Sefton, broadcast each morning, and when he came on the air, 'every miner's wife would turn off her washing machine to stop and listen.'[54] The union bought additional 'air time' when prominent supporters of the strike were in town or when a crucial situation developed.

The radio broadcasts proved to be an effective means of communication. A citizen who was unsympathetic to the union wrote to Premier Hepburn to see whether something could be done to prevent the union speeches from being broadcast over the local radio station. She resented what she considered scoffing at the OPP and the government. She objected to Mrs Shipley being called 'the mouthpiece of the managers.' Such radio speeches had a 'tremendous effect on the listening public. In this case, they keep the strikers pepped up and they put wrong ideas into the minds of the disinterested. This radio station and the press will do anything for money. But we all pay radio licenses and CJKL should not have to allow anything and everything.'[55] She felt the townspeople needed a few 'truths' and wondered if Hepburn could speak on the CIO in a special broadcast.

The union, despite its lack of resources and access to the major newspapers, did a creditable job of presenting its point of view to the local community. To achieve national communication the union had to rely on the Canadian Congress of Labour (CCL), press coverage of the national Kirkland Lake conference, and the work of the Kirkland Lake strike committees across the country. Within the ranks of organized labour these techniques were effective, but they did not reach the general public. To influence national public opinion the union had to rely on factual or sympathetic editorials in the national press. These were few and far between. Organized links with other groups, including the churches and the CCF, meant that portions of the public were provided with information, but otherwise the means of communication were very limited. This was similar to D.C. Masters' analysis of the circumstances of the Winnipeg strikers in 1919, when 'The strike leaders had no ready means of communication with the general public.'[56] Only the clarity of the issue in dispute, the inadequacy of the government's labour policy, and the concern that industrial unrest might disrupt the war effort caused some members of the public to become informed about the strike.

THE 'ANTI-STRIKE' OPPOSITION

The nucleus of the 'anti-strike' organization predated the strike itself. After the IDIC (Industrial Disputes Inquiry Commission) report and management's acceptance of relations with 'employees' committees,' the mine operators began to establish such committees, while at the same time ignoring the union's existence. Some mines held elections for employees' representatives in August 1941, after their employees rejected the IDIC proposals. Others merely set up temporary employees' committees without government supervision, but said that they were still prepared to establish 'permanent' committees[57] in due course.

In some mines these committees began recruitment campaigns,[58] almost always with management support and often at the company's expense. The committees published bulletins and held meetings. They were occasionally privy to information about changes in working conditions before they occurred. Through management, they had learned the rules that the Department of Labour was planning to apply to the strike vote required by PC 7307. The close relationship with management was intended to give the committees an appearance of legitimacy and attract the support of those employees who might wish to publicly identify themselves with management. In fact, the committees merely provided evidence to support the union's charge of 'company unionism.' These committees were seen by management as an acceptable alternative to the union, but it was obvious that they had a hand in their formation and administration, and could influence them.[59] They were not the independent bargaining agent which most of the employees wanted to represent them.

When it became clear that the vast majority of miners were not going to desert the union and that a strike was probably unavoidable, management began preparations to continue operations during the strike.[60] Those miners who had supported the employees' committees and developed a close relationship with the mine operators became the nucleus of the 'loyal' work force during the strike. After the union won the strike vote, the workers' council at Wright-Hargreaves informed McLarty that it was still opposed to union recognition and its members would sleep on the mines' property in order to work if there was a strike. This letter was signed by 163 employees.[61]

Other strike-breakers were brought in before and during the strike. Despite the continued communication with government, and the owners' insistence on the Ottawa meeting, management had no intention of negotiating with the union. As early as 30 October 1941, Reverend E. Long wrote to Dr Mutchmor that 'the mines are determined to operate with their "loyal" workers.'[62] One week before the strike he reported: 'Wright-Hargreaves is planning to keep some 150 or more men right within its property and is provisioning for the siege today. The same will be true of the other mines I imagine.'[63] His estimation was correct. The Teck-Hughes and Macassa mines also had men working and living inside the mine gates. The *Northern Miner* confirmed these activities. Shortly after the union won the strike vote, it reported that within the first week of the strike the larger mines would try to employ 'that large proportion of their men who desire to carry on with the production of gold. For the mines must think of those men who strongly desire to work.'[64] This was, of course, a rationalization. The mine operators' real policy was to break the union.

The union was aware of these company preparations and was determined to undermine them as much as possible. If successful, the companies' plan could weaken the strike and the local. McGuire wrote to J.L. Cohen at the beginning of the strike that 'the ranks are solid here ... the plans are under-way to launch a campaign to smash any attempt at [a] back-to-work move-ment.'[65] These plans included the traditional techniques of mass picketing and sending representatives to talk to the men who continued working in order to convince them not to do so.

When the miners walked off their jobs, the mines' office staffs were work-ing. A small core of non-strikers had remained on the mine properties of some mines, but had not crossed the picket line. The first morning of the strike, in accordance with the Mining Act, the pumpmen and hoistmen had been told by the union to report for work to maintain the mine properties. These workers were turned back by company officials.[66] The decision to con-tinue operations, and the refusal to allow the union maintenance men to work, indicated the firm attitude that the employers had decided to adopt.

On the first day of the strike the union issued a statement that no produc-tion workers were on the job and the mines were entirely shut down. The companies claimed that only two mines were completely closed, two were operating as usual, and the rest were operating on a reduced scale.[67] These conflicting reports continued throughout the strike. The *Globe and Mail* regularly reported management figures. Both sides tried to maintain the morale of their supporters and to demoralize their opponents by the use of statistics.

Periodically the mine operators tried to start a back-to-work movement. In mid December the operators stepped up their activities. Radio announce-ments promised that men would be taken back 'as quickly as conditions permit.' Advertisements of work were placed in the *Timmins Daily Press*. In reply, the union issued a special leaflet that indicated that only 18 per cent of the normal underground work force and 11 per cent of the normal mining machine shifts were working.[68] The union claimed that this was a reduction from the previous week, and told the strikers that some of the strike-breakers had actually joined the strike. The leaflet reminded the strikers that the workers inside were inexperienced, so that it was virtually impossible for the companies to achieve maximum production. Radio broadcasts countered the companies' calls to work. Occasionally the union response included a parade through the town to demonstrate solidarity and encourage those working to 'join the strikers' movement.'[69]

The operators denied importing outside workers during the strike. Instead they explained that 'a very small number of new men have been hired at the mines but these men had come to the gates seeking work and nearly all of them have mining experience.'[70] The mines dismissed what they called the 'so-called' back-to-work movement, saying that men began returning to work the day after the strike began: 'It was started by the men themselves, those men who chose to work. Over 800 men stayed at their jobs when the strike was called. Over 500 have come to work since.'[71] In mid December, management estimated that 1320 hourly paid workers or 40 per cent of the work force was working. The union's figure was 620 workers or approximately 14 per cent. Management also claimed that 240 men had left the employ of the mines permanently. Management denied urging men to work and continued to claim that 'the mines are working because so many men said they wanted to work,' and that more men were going to work daily. The mines repeatedly warned the strikers that the mines could operate on the present partial scale indefinitely, and put pressure on the merchants through the credit bureaus (which they dominated) to refuse to honour union strike-benefit vouchers. The union was forced to curtail expenditures and cut back on the issuance of vouchers and supplies to the strikers. The union believed the operators were trying to incite racial conflicts among the miners in order to undermine their solidarity, since some of the company radio broadcasts had included racial slurs.

It is difficult to get an accurate picture of how many men were on strike, or returned to work, or how many people were imported from outside the vicinity. It appears that the companies were reasonably successful in recruiting, but even the Department of Labour had little accurate data on the question. The companies returned forms at the beginning of the strike that indicated that some companies were better able to continue operations than were others. The Wright-Hargreaves mine reported that as of 18 November 1941, '325 were working and staying on [the] property through special arrangements.'[72] At that mine 725 men were out on strike. The Macassa mine reported that 85 workers, including staff, were carrying on operations. Of a work force of 285 miners, 180 were on strike. The other companies said nothing about their continuing operations. There are indications that, at the beginning of the strike, management figures were exaggerated; however, more workers did return to work as the strike continued. Bob Carlin estimated that at the end of the strike, just over one-third of the total work force was working.[73] The *Labour Gazette*'s figures support this view, although it does not list the same figures as those of the mine operators. Between November 1941 and February 1942, *Labour Gazette* lists a decline in the number of strikers from 2000 to 1500.[74] A police report, dated 20 January 1942, to Ontario's attorney general calculated that 1442 were working at the eight struck mines.[75]

THE INTERVENTION OF THE PROVINCIAL POLICE

The employer's decision to hire strike-breakers increased the likelihood of violence, particularly in a small community like Kirkland Lake, where the conflict between citizens would necessarily be more personal. There were ethnic and class conflicts within the community that created additional tensions. Events leading up to the strike further exacerbated these latent conflicts. The delays during the pre-strike period were acutely frustrating to the miners. In a strike where the union's existence was at stake, the importation of strike-breakers was a direct challenge. To the strikers, the strike-breakers were considered 'scabs' and 'job thieves.'

The law-and-order issue had previously been raised during the Teck-Hughes dispute and during the union's rejuvenated organizing campaign in 1940. In this earlier situation a local police force had been organized on a voluntary basis and consisted of workers, doctors, clerks, and merchants who were to assist in keeping 'law and order' in the town. They were supplied uniforms by the mining companies, and drilled by the local police chief. Their presence unnerved many of the miners who did not know why they were there. Clarie Gillis raised the matter in the House of Commons, and charged that the special police force was company controlled and looked like Hitler's storm troopers.[76]

In 1936, the United States Senate Committee on Education and Labor, headed by Senator Robert LaFollette, Jr, investigated strike-breaking, labour spies, and the use of 'special police.' The committee concluded that these tactics were 'the implements of a labour policy based on the rejection of collective bargaining.'[77] Their presence in a strike situation embittered relations and helped create violence. Moreover, violence itself could serve a purpose. A contemporary of LaFollette's explained the employers' calculations in utilizing anti-union strategies:

It is ever true that the employer has everything to gain by encouraging violence after a strike has begun, since the result is always likely to be direct suppression of the striking element, or at the very least, the marshalling of public opinion against the strikers ... the strikers have everything to lose by rioting and fighting and in fact the union leaders usually go out of their way to avoid any violent display or destruction of property.[78]

This assessment was shared by Stuart Jamieson, who, many years later, conducted a survey of Canadian industrial conflict and concluded that strike-breakers always acted as a 'catalyst for disorder.'[79] He concluded that Canadian employers were just as hostile to collective bargaining as their American counterparts, and ready to use the same means to resist unions. In this respect the Kirkland Lake dispute was not unlike many other disputes that had preceded it. When the OPP marched into Kirkland Lake during the strike, Canadian authorities were responding to the wishes of the mining companies *and* were continuing a long tradition of sanctioning the use of force in labour disputes.[80] The use of outside police, and undisciplined 'special constables' was a provocative element in a number of Canadian and American strikes.[81] Indeed, the tactics adopted by the employers in Kirkland Lake bore a striking resemblance to the 'Mohawk Valley Formula' – a strategy for breaking strikes developed and applied in the United States during the 1930s. Virtually every element of the formula was applied in Kirkland Lake. The summary of the formula in David Saposs' 1938 study of anti-union activity in the United States is a remarkably accurate description of the employer conduct in Kirkland Lake in 1941 (see appendix 2).

Everyone was aware that there might be violence during the Kirkland Lake strike. It was this concern that induced the union to try to restrict access to liquor. The *Northern Miner* reported that the companies were making preparations to protect company property and the men who planned to continue working in the event that there was violence. The paper also commented that the resources belonged to the province and noted that 'the province therefore has a very important stake and possibly would let its mind be known if Ottawa attempted any move which threatened provincial rights in the matter.'[82] Presumably this was a reference to the possibility that the federal government would appoint a controller of the mines, as it had done in response to the National Steel Car strike during the previous summer. The *Northern Miner* much preferred the intervention of Premier Hepburn, whose relationship with the mining interests ensured a sympathetic response.

The townspeople suspected that the provincial government might intervene in the strike. As early as October 1941, Reverend E. Long wrote, 'It is ... rumoured that Hepburn is waiting to jump into the local situation in defense of his friends against the CIO.'[83] Attorney General Conant visited Kirkland Lake on Sunday, 2 November, on an 'undisclosed mission.' Reverend Long suggested Hepburn might be thinking of calling a quick election on an 'oust the CIO and embarrass King' platform. Hepburn did not run an election campaign against the CIO as he had in 1937, but federal government politicians believed that this was a possibility. This belief influenced their 'hands off' policy towards the Kirkland Lake strike.[84]

The companies knew that they had strong allies at Queen's Park, as was later demonstrated. For example, during the strike the secretary of the Wright-Hargreaves Workers' Council, Clarence Chanonhouse, requested and received from Attorney General Conant a special contingent of Ontario Provincial Police to protect the wives and families and property of his members. He represented 383 workers who were living on the mine property and working in the mine. Conant replied: 'All persons who desire to work will be protected in the exercise of their right to do so and violence of any kind will be prevented by every possible means ... Those who break the law will be vigorously prosecuted and if convicted the courts will be urged to impose substantial penalties.[85] Conant had shown no such zeal in enforcing section 502A of the Criminal Code which, since 1939, made it an offence for employers to discharge or discriminate against an employee because of his support for a trade union.

Reverend Long, who was privy to a great deal of information in the community, realized that the continued operation of the mines would inevitably lead to trouble. He noted that the companies planned to hire guards to patrol their power lines, although the union was not interested in pursuing a policy of sabotage, and there was no evidence whatsoever of this possibility.[86]

When the strike began, the Kirkland Lake police force, consisting of a few special constables under the command of Chief Constable Pinegar, were responsible for maintaining order. Picketing was peaceful. At the beginning of the strike, Reeve Carter acknowledged that the picketing was orderly, but suggested that the local police should obtain additional assistance if necessary from citizens trained in police work who could be sworn in as special constables. By 20 November no violence had been officially reported, but the accusations began. N. Bryson, president of the Lake Shore Workers' Council, protested to Reeve Carter that his members had been 'molested' while walking to a meeting. The union claimed that the reports of intimidation were 'incorrect' and 'company propaganda.' Carter made a special radio broadcast to ask both sides to use restraint and co-operate with the police in order to preserve law and order 'without recourse to outside assistance.'[87] There had been no serious incidents and Carter hoped that this state of affairs would continue.

According to one newspaper, some of the 'boys' at one of the mines had decided to conduct a percussion concert late one evening for the benefit of workers bunked in a company building on company property close to the fence. Outside the fence pickets banged on oil drums to keep the sleeping men awake: 'Such hijinks have characterized most of the violence so far.'[88] The spirit of the strikers was high, the lines were orderly, and the union continued to respond that the operators' charges of intimidation were pure nonsense. On 21 November the *Globe and Mail* noted that the union and the operators differed both as to the number of men working in the mines and the conduct of the pickets.

On 22 November 1941, the operators charged that workers were being attacked and threatened, that there had been beatings, and that 'acts of lawlessness were becoming frequent and were taking on a serious aspect.' The companies contended that sixteen workers who were entering the Bidgood mine had been stopped by fifty pickets and two men had been attacked. Despite such 'intimidation,' management reported that men were going through the picket lines in increasing numbers. However, they charged that 'peaceful picketing as recognized by law does not exist in Kirkland Lake,' and had not from the beginning of the strike. The progress of cars going through the picket lines was allegedly being hindered by pickets, and tacks and nails were thrown on the roadways to the mines.[89]

In a long letter to Reeve Carter, dated 22 November 1941, the union asserted that the operators were distorting the situation. Faced with the unity and determination of the Kirkland Lake miners in the strike situation and 'finding themselves unable to confuse the workers, unable to get across their distorted and false message and unable to shake the solidarity of Kirkland Lake workers ... the operators seek now to institute a reign of terror and hope by this means and by a strong show of provocative force to destroy the present orderly state of the strike and substitute it with disorder and violence of their own making.'[90] The union carefully explained each of the incidents of which the company had complained. The most serious was a fist fight in a tavern between two men who had had too much to drink – not an unusual event in a mining town even in normal times. There had been no charges, no arrests, and no actual violence, although there undoubtedly had been 'name calling' by both sides. The union repeated its offer to supply men of good standing in the community to act as deputies. The companies' charges were portrayed as an attempt to create unrest and undermine the union's credibility.[91] The union considered the request for outside police an attempt to further the interests of the companies that could only create difficulties. Outsiders would not understand the local situation and personalities and would be more likely to resort to force in the first instance. Their very presence would be a provocation.

The police continued to receive management complaints of intimidation at the mine gates, on streets, and at workers' homes. Despite the union's view that nothing untoward had occurred, Chief Constable Pinegar advised Reeve Carter to request the aid of the provincial police. His portrayal of the situation was completely different from that of the union. Pinegar alleged that incidents of intimidation had increased and claimed that there was 'definite proof of the use of organized bands of thugs,'[92] although he did not outline the nature of that 'proof,' or indicate why no arrests or charges had been made. He was convinced, however, that the union staff did not intend to carry on 'at least a partly clean fight.'[93]

The terminology used by Pinegar was virtually identical to that subsequently contained in the OPP Annual Report. The report labelled Tom McGuire a 'union organizer, leader and international agitator,' and alleged that before the OPP arrived, 'loyal workers' were being 'shanghaied' by strikers and forced to proceed to the union office and become union members. Neither police force was neutral in the strike: their role was not to maintain the peace but to protect the property of the mines and the 'loyal workers' who wanted to work. The local police had been flexible in maintaining the peace, for they understood the dynamics of the situation in their own community. The provincial police acted with rigid military discipline throughout the dispute.

On the same day that both Pinegar and the union had addressed their views to Reeve Carter, M.W. Summerhayes, manager of the Wright-Hargreaves mine, wired Premier Hepburn to inform him of a number of incidents between picketers and strike-breakers. He accused the union of threats, intimidation, and assault, and claimed, 'the local police chief admitted to me this morning that his force was utterly inadequate to handle the situation and would welcome assistance.' He urged 'immediate action on [the] part of provincial authorities to prevent bodily harm to employees who want to work.'[94]

There is no doubt that the Ontario government was ready and willing to act. It was waiting only for an opening. Reeve Carter unwittingly provided it for them. He wrote to Attorney General Conant, enclosing Pinegar's letter, saying, 'We find it physically impossible to carry on with our existing force who have been working 16 hour shifts. As I stated last week when I was in Toronto we hoped to handle the situation but the time has now come when we must seek your aid. Chief Pinegar feels the situation would require 30 men to handle it and give some relief to his own men.'[95] The Ontario government acted immediately. The Ontario Cabinet met on the morning of 24 November 1941 in a special meeting called to discuss the Kirkland Lake situation. After this meeting it was announced that the Teck Township Council had asked the province to police the township, 'with complete jurisdiction over the present municipal police force.' Attorney General Conant announced 'law and order will be maintained at all costs.' The 'men at Kirkland Lake and elsewhere who want to work will be adequately protected in the exercise of their right to do so. Property will also be protected from illegal activities.' In a similar vein, Hepburn stated that 180 Ontario Provincial Police would be sent to Kirkland Lake and that their numbers would be augmented if necessary. He declared that there would be law and order 'at all costs ... if we have to recruit an army.'[96]

Hepburn justified his government's position on the basis of the federal government's ruling that provincial authorities were responsible for law and order in Kirkland Lake. Government action was justified by Pinegar's report and the long letter from M.W. Summerhayes. Hepburn told a press conference, 'We have direct charges that women have been intimidated, men have been dragged out of cars and assaulted and the King's highway blocked.' He continued that Ontario's jails and reformatories 'are just yawning for John L. Lewis' paid organizers.'[97] His government's policy, he stressed, had not altered since the 1937 election 'when it was endorsed by the people in regard to its opposition to the CIO.' He believed that the purpose of sending in the police was to enforce law and order, stop intimidation, and 'reserve for the men who want to work, the right to work.' He also announced that the government was sending a special Crown Prosecutor, W.B. Common, KC, to the district, so that those charged with violence and intimidation would be tried immediately.

Following these government announcements, Conant wired Reeve Carter. He referred to Pinegar's phrase, 'organized band of thugs and mobile units of intimidators,' and stated that provincial police were proceeding to Kirkland Lake immediately '*in numbers considered necessary* by Commissioner Stringer who ... will be in command of them.' Carter's request for thirty police was ignored and control of the number of police was removed from the local authorities and placed in the hands of the *provincial* commissioner of police. This action was taken because of the statements of the local police chief, who (notwithstanding his statement) had himself estimated that thirty OPP would be adequate to handle the situation. Consequently, Commissioner Stringer, 1 staff inspector, 3 district inspectors, 6 sergeants, 4 corporals, and 168 provincial constables 'took over the policing of the strike-affected areas of the Township of Teck ... previously under the control of Teck Township Police Force.'[98]

When informed by the province that the government was sending police in large numbers, Reeve Carter protested that thirty police were all that Pinegar thought necessary to augment the local force of fourteen. He insisted that on 15 November in an interview with Attorney General Conant, an understanding had been reached to the effect that the provincial government would assist the local police only to the extent considered necessary by the *local* Chief Constable Pinegar. Carter believed that the 'excessive' force was provocative. Conant denied that he had any understanding with Carter, and pointed to the facts that Carter had disclosed in his own correspondence. The government regarded the situation in Kirkland Lake as 'serious to the nation and the Province as well as to your municipality.' Apparently the local officials on the scene simply did not appreciate the gravity of the situation.

On 24 November, the council chamber was filled by several hundred miners who had gathered to attend the special council meeting called to discuss the arrival of the provincial police. The majority of council considered this police intervention a reflection on the good character of the community and 'highly provocative in the present situation.'[99] There was a prolonged discussion of Reeve Carter's role, which enabled the police to legally intervene, and while the council was divided, ultimately it censured Carter for requesting additional police without first informing the council. The majority were seriously concerned that the intrusion of large numbers of police would merely inflame the situation. Pinegar reported that he had made nine arrests – none of which involved serious offences. The council passed a number of resolutions calling for the withdrawal of the OPP, requesting the periodic closing of liquor outlets, and suggesting that the provincial government investigate the possibility of arbitration. Hepburn rejected all of these proposals, which he charged were motivated by the up-coming municipal election.[100]

While the municipal council was divided, the local police commission (composed of Reeve Carter, Magistrate Atkinson, and Judge G.H. Hayward) endorsed Pinegar's position. This decision ended the debate about the outside police, although not the bitterness their intervention had created. The union's position as stated by the local's president, William Simpson, remained the same. He charged that 'the operators in Kirkland Lake and their associate interests are engaged in a vicious campaign to create the false impression that conditions in Kirkland Lake ... require the importation of outside police units.' Even Commissioner Stringer stated shortly after his arrival that, except for three incidents, which occurred while the men were proceeding 'lawfully' to their work, 'I have found the conduct of the pickets quite commendable.'[101]

The response of the general public to Hepburn's action was mixed. D.G.H. Wright, a mining geologist and engineer whom Hepburn knew personally, congratulated Hepburn on sending such a large number of police, since 'an ounce of prevention was worth a pound of cure.'[102] Other correspondents disagreed. One man swore he would never vote Liberal again because of Hepburn's 'threatening' manner towards the miners. A woman wrote that the operators were trying to discredit the miners in the public eye, that undoubtedly there were 'hotheads' among the miners but that the press reports exaggerated incidents. The executive secretary of the Vancouver Labour Council protested the importation of special police, stating that the union had tried to get a settlement by conciliation for months, and that the owners were responsible for the strike. Another trade unionist believed the number of police was provocative and this 'is again an attempt of your government to interfere in the struggle of Canadian workers to organize in trade unions of their own choice.'[103] The MPP from New Liskeard, W.G. Nixon, congratulated Hepburn and assured him that local businessmen supported his action.[104]

There is no question that many members of the middle-class community of Kirkland Lake were relieved that the OPP had arrived to 'protect' them. They believed the atrocity stories and some of them felt that the town was on the verge of civil war. One correspondent to Hepburn concluded a very distressed letter with, 'If this CIO is not smashed right here, it will continue to spread and perhaps control the whole country. That at least seems to be their ambition.'[105] Another worried citizen was very pleased with Hepburn's action but complimented him by the use of an analogy not altogether flattering to the premier: 'Hitler would know what to do with [John L.] Lewis and his gang. They would be stuck up against a wall and shot for making trouble like they do in wartime. We need more men like you.'[106]

After the OPP arrived, the number of charges against the strikers increased. One of the most serious involved two cases of dynamite that were found on the doorstep of two non-striking miners. The dynamite was not fused and there was, therefore, little actual danger, but the incident was disquieting. The union denied any responsibility and asserted that the incident was a clever ploy to discredit the union. 'This same trick of planting dynamite and blaming the union for it is as old as the hills but the operators are not above using any trick to defeat the union.' The union argued the only place dynamite could come from was behind the mine gates. There was no evidence linking the union to the incident. No charges were ever laid. The origin of the dynamite was never determined. However, Attorney General Conant concluded that 'the *facts* from Kirkland Lake regarding the use of dynamite indicate that the CIO are up to their old and well known tactics – terrorism. The suggestion that the dynamite was planted by persons other than the CIO agents is too absurd for comment.'[107] Such statements did little to inspire confidence that the police would impartially enforce 'law and order.'

Charges against the strikers contributed to the anxiety of union leaders and strikers, and to expense for the union. The miners could not afford to hire their own lawyers so the union had to provide counsel for them. J.L. Cohen believed that the police were making it difficult for union men to lay charges, while actively encouraging strike-breakers to complain.[108] Of even more concern was the apparently widely held police view that the strikers were revolutionary terrorists committed to violence, sabotage, and subversion to gain their objective, which was to 'take over' the mines. The OPP Annual Report stated: 'It was the intention of the union to draw a cordon so tightly around the affected mines that no person would have been able to go in or out and then by sabotage and terrorism force the operators to accede to their demands.'[109] The union of course had no such intentions; and these paramilitary appraisals reveal more about the biases of the OPP than they do about the union's strategy. The language of the report was similar to the rhetoric of the mine managers and the provincial government spokesmen.[110] None of them regarded the miners' objective as legitimate. The union's demands were a threat to the employers' authority and a CIO success would alter the *status quo*, but there was no question of sabotage or a 'take over' of private industry. However, neither the mine owners nor the provincial government nor the police made this distinction.

The actions of the police contributed to the already uncomfortable situation of the strikers. The strikers had pitched tents near the picket lines and the various mine gates so they could warm themselves and make coffee. Tents were located at Government and Goldthorpe roads leading to the Macassa mine, at the junction of Larder Lake Road and the road to Bidgood-Kirkland mine. The strikers had also built shacks of railroad ties and lumber on the T & NO Railway, traversing the north side of mine property. The police regarded these shields from the cold as being placed at 'strategic points' and ordered that they be dismantled, and all the stoves, axes, and cordwood found inside them removed. A striker's wife remembered that '[the pickets] would huddle in there at night. After all, it was 40 below zero, and they would get in there. So they [the police] tore down the shelters which were made of garbage cans and what not, so the guys had to stand there.'[111]

The police also banned mass picketing. Pickets were reduced to six or seven at each of the main entrances of the eight strike-bound mines. The official rationale is interesting in that it reveals the police attitude to immigrant workers who worked in the mines: 'It was felt that such action was necessary as a high percentage of the men on strike were of foreign or alien birth, and when in numbers sufficiently strong to intimidate the loyal workers or their families were most ruthless in their conduct.'[112] The purpose of the mass picketing was to demonstrate solidarity and make it uncomfortable for strike-breakers to cross the picket line. The removal of most of the pickets made picketing ineffective.

Armed guards were assigned to patrol the Ontario Hydro transformer station on Government Road, which supplied power both to the mines and to the community. The police claimed they had heard rumours that an attack upon it was being planned. As the source of the rumour was a company official, nothing happened![113]

From the beginning, neither the strikers nor the members of the middle class in Kirkland Lake (who were happy with their presence) regarded the police as 'neutral' figures merely keeping the peace. On 5 December, Attorney General Conant went to Kirkland Lake for a one-day inspection of the situation. He concluded that the number of men returning to work indicated that there was confidence in the community that 'every man who wants to work will be adequately protected in his right to do so.' When the association of clerical, supervisory, and technical employees of the Kirkland Lake mines commended Conant for his stand on police protection, the attorney general replied that he was pleased that a substantial portion of men had returned to work 'and that more are returning everyday.'[114] He reiterated that it was his duty to provide law and order in the province and to protect all those disposed to exercise their right to return to work. In his mind the two 'duties' were synonymous.

To label the activities of the police harassment is perhaps too strong, but there is no doubt that the individual policemen had little sympathy for the strikers and were zealous in enforcing law and order. Nor did they hesitate to indicate their views concerning the union,[115] or their belief that picketing was an illegitimate activity.[116] Many of the criminal charges were laid by the police; there was a high number of dismissals. Assault charges were largely confined to jostling or verbal abuse of strike-breakers – bothersome no doubt, but not examples of serious violence. The number of cases that were dismissed and the lurid descriptions of relatively minor incidents suggest that the criminal character of the strikers' conduct was greatly exaggerated.

The actual extent of unlawful conduct during the strike is difficult to determine. The *Globe and Mail* reported 'many instances of violence and intimidation' by the CIO and 'numerous complaints' against strikers. Curiously, these complaints did not result in a large number of criminal prosecutions, despite the number of police on the scene. Bob Carlin minimized the incidents that did take place. He claimed, 'We paid $12 000 in petty fines; today you wouldn't be fined for them at all; they don't lower a Court to entertain them.'[117]

Some violence did occur; some was initiated by strikers and some was the result of actions by strike-breakers or police. Generally union discipline of the strikers was maintained and violence was minimal. Stuart Jamieson's depiction of the 'violent overtones' of the Kirkland Lake strike[118] is inaccurate, for although there were a substantial number of charges laid, these were mostly minor and for the most part resulted in small fines or dismissals. At the end of the strike even Attorney General Conant admitted to a mine manager that he was 'pleased that in the final result there were no serious breaches of the law and comparatively little property and physical damage.'[119] Most of the problems were associated with the use of strike-breakers. 'There was trouble. Some of the people were beaten up and some of the strikers, and you couldn't blame them, were stopping scabs out in the back lots ...'[120] This was not violence organized by the union – as management, the police, Mitchell Hepburn, and the *Globe and Mail* believed. It was action taken by individuals who were frustrated by their forced idleness, the hostility of the public and press, and the concern that others were taking their jobs. Such acts were discouraged by the union.

J.L. Cohen was defence counsel for most of the union members charged with criminal offences. An examination of his files gives some indication of the seriousness of the situation. Information is available on thirty-one defendants. Of these, nineteen were acquitted outright and in one other case, the trial was adjourned and the charges dropped. In one case leave to appeal was granted. Four cases resulted in small fines of between five and twenty-five dollars. The other cases resulted in more serious convictions. One conviction was for property damage and involved two strikers who were charged with damaging six storm windows, nine windows, one door, and six blinds on the house of a man living and working at Teck-Hughes mine. One striker was convicted in June 1942; the union paid $80 for repairs. Three jail sentences of from thirty to ninety days were handed out, and one case resulted in a sentence of $100 fine or time in jail. The record of the union members was not totally clear, but the charges and the convictions do not reflect the diabolical character attributed to their acts by such newspapers as the *Globe and Mail*.

One important case that was dismissed involved Local 240's vice-president, Joe Rankin. He had been arrested for intimidation when two strike-breakers charged that Rankin had threatened them in order to keep them away from the Teck-Hughes mine. Both complainants had criminal records. One had been convicted previously of theft, and one was a deserter from the army. Rankin's case was dismissed as it turned out that one complainant had himself been jailed the previous night for theft. This incident prompted the CCL secretary-treasurer, Pat Conroy, to protest the arbitrary arrests of strikers in Kirkland Lake, and especially that of Joe Rankin.

It is highly unlikely that the union desired violence, which could only work to its disadvantage. Initially it offered to appoint deputies from within the ranks of the strikers to aid the small local police force to maintain order. It had tried to get the local bars closed for the duration of the strike in order to prevent drunkenness among its members. The local itself organized activities and entertainment to prevent idleness and keep people constructively occupied. It had consistently tried to demonstrate its responsibility in order to meet the employers' claim that it was 'irresponsible.' Prior to the strike, it had complied with all of the orders-in-council affecting labour relations even though it resented the delay. After the strike it paid its debts and assisted the miners to get rehired or to find new jobs elsewhere.

What violence did occur was spontaneous and unorganized. It was probably unavoidable under the circumstances. As Dr Coburn of the United Church wrote to his beleaguered local ministers, 'Regarding the matter of violence. Any strike involves more or less strife with mutual recrimination. When you consider, however, the large number of men involved in the strike, I think the actual cases of violence as dealt with by police is small, and surely we can rely upon Mr. Hepburn's 180 huskies to see that no cases were overlooked.'[121] The exaggeration of the nature and amount of violence both by management and by the press had several effects. Initially it gave the only too willing provincial government a reason to send in police. Later the reports and charges of intimidation helped discredit the miners and caused them to lose some public sympathy. It also served to draw attention away from the primary reason for the strike – the recognition of the union and the rights of the miners to bargain collectively with their employers.

The behaviour of the police proved to be intimidating to the strikers. Every morning at eight o'clock the entire contingent of police would parade through the town, apparently for no specific reason except as a show of force. The OPP Annual Report defended the need for police in Kirkland Lake and their effectiveness once there, saying that 'the sending in of sufficient numbers of all ranks of this force had a salutary effect on strikers and their sympathizers.'[122] This is debatable. The continued presence of great numbers of police was bitterly resented. The peace was kept at great expense to future good relations in Kirkland Lake. The role of the police was seen by the miners as supporting the position of the mine operators and helping them to run the mines with 'scabs.'

The police and the provincial government did not see themselves as neutral, peacekeeping forces. They were openly sympathetic to the strikebreakers who had 'a right to work,' but not to strikers who were picketing to protect their jobs. The police perceived the union as a centrally directed, paramilitary organization engaged in strategic operations against the mining companies. This lack of sympathy for the strikers and complete misunderstanding of the union created bitterness and alienation. The union had sought to comply with the law even though it considered the legislation inadequate and unfair. Now it appeared that the provincial government and the police were in league against it. They were accused of plotting a take-over of industry when all that was sought was a face-to-face meeting with management.

Management charges of intimidation of 'loyal workers' and the preponderance of charges and arrests of strikers – and not of strike-breakers – created the impression that all the violence was committed by the union men. This impression influenced some people in the local community. The United Church ministers, who were among the most sympathetic to the union, for a short time began to doubt its responsibility, and started to echo arguments of the mine operators. The fact that the strikers were blamed for the violence worsened relations between the two sides. The fact that most of the charges were dismissed was not reported.

The early intervention of the police in the strike did have the effect of assisting the mine operators. At the end of the strike Hepburn received several letters of thanks from mine operators and the editors of pro-management northern newspapers. M.W. Hotchkin, general manager of Wright-Hargreaves mine, wrote to Attorney General Conant and Premier Hepburn on behalf of the Kirkland Lake Operators' Committee: 'Without the assistance of the Provincial Police, we are sure that the situation would eventually have been completely out of control. The measure of *support* afforded by the Police *enabled many of our men to continue to work and hastened the end of the strike*.'[123] He thanked Hepburn for his interest in the situation and concluded that 'there is no doubt in our minds that the final outcome might otherwise have been entirely different.'[124] What Mitchell Hepburn had failed to accomplish in Oshawa in 1937 he was able to do in Kirkland Lake in 1941. The mine operators, unlike the management at General Motors, were as obdurate as Hepburn. The tactic of sending in a large number of provincial police as soon as a strike broke out, the use of police to intimidate strikers, to arrest their leaders, and to assist strike-breakers through picket lines would continue to be used in other strikes during the wartime and immediate post-war period as, for example, in 1949 at Asbestos, Quebec.[125]

THE SEARCH FOR ALLIES

While the union's primary concern was to maintain solidarity, communicate with the citizens, and cope with the police in Kirkland Lake, it also sought to establish links with sympathetic groups across the country. Churchmen, such as Mutchmor and Coburn, were helpful allies, and the CCF provided a political voice, but the principal support had to come from the labour movement itself.

One week after the strike began, J.L. Cohen visited Kirkland Lake and subsequently wrote to Mine Mill President Reid Robinson: 'It appeared to me that no concrete steps had been taken to gear into the whole issue the support of the labour movement generally and particularly the Congress of Labour membership in Canada, to say nothing of the sister locals and affiliated unions in the United States.'[126] He warned that this situation had to change if the strike was to be won. It was decided that it would be useful to hold a national conference in Kirkland Lake that would be attended by the national representatives of leading unions affiliated to the Canadian Congress of Labour.

Telegrams were hurriedly sent out in an effort to organize this conference within a week. It took place on 29 November 1941. The objective was to transform the local strike into a national labour issue and to make it a test case for the government's labour policy. The demand for a 'proper settlement' of the dispute was linked to the general demand for a full 'partnership' in the war effort. A statement was issued to this effect, and was included in a telegram to Prime Minister Mackenzie King. The government was informed that an 'emergency conference' representing a total of over 100 000 organized workers had unanimously endorsed the Kirkland Lake strike. The labour movement pledged its complete support – financial and organizational – since the issues were too fundamental to ignore:

Collective bargaining is the recognition of the right and need of labour to be admitted to partnership in the war effort and war management. The mine operators of Kirkland Lake, and their associates, by precipitation of the Kirkland Lake strike, by the contempt shown by them to the proceedings and recommendations of the Dominion government Conciliation Board and by their undemocratic refusal to accept any proposal of arbitration, have deliberately challenged labour's right to war partnership. Canadian labour must accept that challenge.

Recognition of labour as a partner in the war secures that national unity and heightened war morale of the people which alone can produce the maximum war effort essential for success and victory. It is also the most effective means of exemplifying to the workers of Canada the fundamental war purpose of securing democracy and smashing Hitlerism. This can only be accomplished by permitting and encouraging the formation of unions of the workers' choice so that by these means they can play their proper part in industry and in society.[127]

The statement urged the prime minister to enact legislation that would facilitate collective bargaining and recognize the right of labour to a proper place in industry and the war effort. It even appealed to Mackenzie King on the basis of national unity and requested a conference in Ottawa to discuss ways to avert a serious national situation. Nineteen leaders signed the telegram and copies of it went to thirty-eight union leaders.

King did not respond to this statement until 6 December. He refused to intervene in the strike – a reply that McGuire considered 'a continuance of the government's policy of withholding any actual aid or assistance in respect to the issue of collective bargaining.'[128]

Following the National Kirkland Lake Conference, efforts were made to secure organizational and financial support from other locals. Twenty-nine special committees were established. Much was accomplished in one week, and the amount of money raised demonstrated the interest in the miners' strike. The most significant initial financial support was offered by District no. 26 of the United Mine Workers' union in Nova Scotia. It pledged fifty cents per member per week. Members of District no. 18 (Calgary) of the same union assessed themselves one shift's pay a month per member, for a total payment to the strike fund of over $10 000 per week. The SWOC national office advanced the strikers $20 000.

A number of individual SWOC locals sent cheques – the most significant amount being $1000 from the recently chartered local in Algoma. This amount represented one day's pay per member. The Toronto ACWA (Amalgamated Clothing Workers of America) locals assessed themselves one day's pay, which amounted to a $5000 contribution. The United Automobile Workers' (UAW) decision took longer because of ongoing negotiations at Ford in Windsor; however, on their own initiative, the Oshawa local sent $2000 and the Windsor local sent $1000. The formula of contributing one day's pay was actively promoted throughout the country by the national headquarters of CCL unions and the recently organized Kirkland Lake strike committees.[129]

In December, Reid Robinson urged all locals and ladies' auxiliaries of the IUMMSW to give financial support to the miners in Kirkland Lake, for 'victory in the Kirkland Lake struggle will mean a death blow to the Rockefeller company union plan which is threatening to spread throughout Canada and the United States.'[130] He believed that a victory in Kirkland Lake would facilitate the future organization of 40 000 Canadian workers. This view was shared by the mine operators who saw the Kirkland Lake struggle as determining the future of union organization in the Canadian mining industry. Many American locals of Mine Mill and other CIO unions sent financial contributions.

By the end of the twelve-week strike, expenditures amounted to $145 000, of which over 80 per cent was for food. Mine Mill contributed $32 286. Some $23 027 was collected from the international office; the rest came from donations from locals in Canada and the United States. The United Mine Workers sent $36 410 of which $28 290 was from the eastern district headquarters. The steelworkers contributed $30 366 of which $25 000 came from their international office. The UAW gave $12 534 of which $6500 was sent from their international headquarters. Amalgamated Clothing Workers contributed $5200 of which $5000 was from the Canadian locals and $200 from their Toronto Joint Board. Strike committees set up in various cities collected an additional $5162. More than half of this was contributed in Vancouver. Labour councils gave $1025. It is indicative of the widespread sympathy for the strike that the Halifax Trades Council – an AFL (American Federation of Labor) affiliate – made a contribution, as did other AFL unions.[131] The Toronto Labour Council at the time had a policy of contributing $10 to strikes that requested financial aid. Despite this policy it contributed, on several occasions, to the strike *and* chose its president to represent it on the local Kirkland Lake strike committee. Despite these efforts the debts mounted.[132]

Local Kirkland Lake strike committees were established across the country in areas where there were substantial groupings of workers. Their function was to organize mass support for the strike and stress the issue of union recognition. As J.L. Cohen saw it, 'These committees will constitute the nucleus of a national apparatus which, both from the standpoint of money and securing organizational support, will enable complete co-ordination of activity.'[133] As early as 9 December the Toronto District Kirkland Lake Strike Committee convened in order to solicit support for the miners' objectives and strike fund. That committee and several others from well-populated areas were expected 'to give the lead to others in the country.' Ultimately there was a Kirkland Lake committee in every major industrial centre in the country. Unions demonstrated their solidarity with the miners by passing resolutions of support in their local meetings and informing the provincial and federal governments of their position. Both governments were bombarded with mail throughout the strike.[134] For the most part, such communications were simply acknowledged by the government.

Management's public response to the Kirkland Lake conference and to the labour movement's support of the miners' strike was to charge that the CIO was trying to place responsibility for the dispute on the government. Employers viewed the co-ordinated action by organized labour as 'lawlessness,' contrary to national unity and the war effort, a veiled threat to the government,[135] proof of irresponsibility of unions, and justification for their refusal to engage in collective bargaining.

In order to co-ordinate the district Kirkland Lake committees, a conference was held in Ottawa on 13 December 1941 and a National Kirkland Lake Strike Committee was established. This conference followed the Kirkland Lake conference by two weeks and, as expected, it was attended by those leaders who had previously signed the Kirkland Lake statement, as well as by representatives of various district committees. The national committee included all of the principal leaders of the CCL.[136] The unions hoped that a national strike committee would emphasize the national character of the issues in the Kirkland Lake struggle, and its presence in Ottawa would make it possible to pressure the government more effectively. Throughout the strike this committee issued strike bulletins to all CCL-affiliated locals and to labour councils across Canada.

At the Ottawa conference, Tom McGuire and William Simpson presented an encouraging report. Efforts by the mine operators to organize a back-to-work movement had been largely unsuccessful. The union had received generous contributions and plans had been made to ensure nation-wide support for the strike-relief fund. The report optimistically suggested 'there would be no difficulty in meeting the needs of the strikers and their families as long as help was required.' At the conclusion of the conference, there was another request for a meeting with the prime minister to consider the Kirkland Lake situation and the government's labour policy.

Between these two conferences, the United States entered the war. As Cohen noted at the time, 'The impact on everything will ... be tremendous.'[137] He was uncertain just what the effect would be. Dr Mutchmor believed this international development would distract the government's attention away from the issue of collective bargaining.

The Kirkland Lake committee structure functioned well throughout the strike, but it was the support of *individual* unions that captured the attention of the press. In the midst of the strike, a conference of twenty-five SWOC delegates, representing 15 000 workers, 'wholeheartedly' endorsed the actions of the Kirkland Lake miners and pledged as much support as was necessary to assure their success. The delegates called on the prime minister to deny war contracts to employers who refused to recognize and negotiate with the unions freely chosen by a majority of their employees. Charlie Millard warned that if the government would not cope with 'this anti-labour and anti-social group of selfish employers' and did not act quickly in the Kirkland Lake dispute, it was certain that trouble 'of serious proportions' would develop in the steel industry of Canada, and in other industries as well. The *Globe and Mail* commented that Millard's statement was a threat that demonstrated the CIO's tyranny and irresponsibility, and 'implied treachery to the country, industry and honest labour.'[138]

Some support came from the TLC (Trades and Labour Congress) even though the CCL was a rival federation that was still in competition with the TLC. The leaders of the TLC realized that the principles at stake in Kirkland Lake were shared by all trade unionists. One week after the strike began, TLC president, Tom Moore, warned that labour unrest in Canada was widespread, and appealed to the mine owners at Kirkland Lake to review their position and not allow personal prejudices to interfere with their judgment.[139] Throughout the strike some TLC locals gave financial support and encouragement to the Kirkland Lake miners.

As a result of these conferences, and the activities of the Kirkland Lake committees, interest in the dispute spread beyond labour circles to other groups in general sympathy with the objectives of organized labour. During the Depression and early war years, many churchmen had become interested in labour issues and had sought to learn more about the problems of working people and the trade union movement. This concern increased when it became apparent not only that workers were vulnerable to exploitation, but also that resolution of their grievances would be important for the war effort. Some churchmen became actively involved in legislature lobbying; others campaigned on behalf of civil rights when labour leaders were interned in 1941 and 1942.

Within the churches there was a recognition that industrialization had brutalized workers and condemned many working-class people to a drab, regimented, alienated existence. Employees were treated as commodities in the production process and were often regarded as such by employers. It was recognized that the Christian churches had not come to terms with these changes and the social problems that had been created by this dehumanizing process. As in the earlier period of the 'Social Gospel' movement, it was recognized that a wide gulf had been created between the churches and labour. As one minister put it: 'If we are ever to have a decent order of society that gulf must be bridged. If the church were to take any action that would assist those who are trying to thwart the legitimate aspirations of labour, it would only widen and deepen that gulf, and render its bridge more difficult in the days to come.'[140] In this spirit a multi-denominational church committee approached the Toronto and District Labour Council regarding a joint labour/church conference. The Christian Labour Forum, composed of leading Protestant churchmen and trade unionists was formed in 1940. It supported the Kirkland Lake strike and demanded that the government make the recommendations of the conciliation report mandatory.[141]

The United Church, through its social instrument, the Board of Evangelism and Social Service, was the most actively concerned with the problems of organized labour, and it became deeply involved in supporting the Kirkland Lake strike. The activities of its director, J.R. Mutchmor, and his assistant, J.D. Coburn, went well beyond writing encouraging letters to the local ministers in Kirkland Lake. They kept abreast of current labour disputes and the issues involved. Their department published educational pamphlets such as 'Industrial Organization and Collective Bargaining' for discussion in various United Church presbyteries. The United Church itself took policy positions on labour issues as early as 1936.

In its annual report of 1941, the board recommended the acceptance of collective bargaining and quoted the words of American Supreme Court Justice Hughes, who in 1937 wrote the decision that ruled the Wagner Act constitutional: 'The right of employees to self-organization and to select representatives of their own choosing for collective bargaining is a fundamental right.'[142]

In 1942 the board reviewed the relations between the church and labour and its positive stand regarding collective bargaining. The board was gravely concerned about the right of labour to organize, 'and had viewed with alarm the curtailment of this right, through lack of a clearly stated federal policy in favour of collective bargaining. It has followed with interest the findings of conciliation boards in general and the finding of the Kirkland Lake Board in particular.'[143] In 1941 the board had commended the government upon the passage of PC 2685 and PC 7440 and had been complimentary about labour minister McLarty's performance. As a result of the church's involvement in the dispute in Kirkland Lake, the board adopted a position on collective bargaining that was virtually identical to that of the two trade union congresses. The board noted in 1942 that 'owing to Mr. McLarty's failure to give strong leadership in industrial problems, the Toronto Conference of Evangelism and Social Service Committee ... sharply criticized Mr. McLarty.'[144]

Doctors Mutchmor and Coburn put their principles into practice throughout the Kirkland Lake dispute. Their efforts influenced the passage of the resolution of the Toronto West Presbytery. These actions were publicized by the *Toronto Star*,[145] which was one of the few large daily newspapers sympathetic to the miners. This publicity counteracted the constant anti-labour attacks appearing in newspapers like the *Globe and Mail*. Their criticism of McLarty and his role in the strike may have influenced the government to choose a new labour minister.[146] Coburn and Mutchmor wrote to the government often urging intervention in the strike. They received a fuller response than most other members of the public, who usually received a 'form-letter' reply.

Mutchmor and Coburn requested an interview with the prime minister or the minister of labour so they might personally express their views. It took Dr Coburn two weeks to contact the new minister of labour, Humphrey Mitchell, and to request a meeting with the church leaders and representatives of the Fellowship for a Christian Social Order (FCSO). His persistence finally succeeded, and on 1 January 1942, Mitchell wired Coburn that he would see the delegation in Ottawa on the following afternoon. A few members of the delegation were unable to attend on such short notice; however, seven ministers, representing the FCSO and the Christian Labour Forum met with labour minister Mitchell for half an hour. They were well received, and as Coburn later reported to his friend, Joseph Atkinson, the publisher of the *Toronto Star*: '[Mitchell did] not utter one word in defence of the attitude of the operators at Kirkland Lake.'[147] Mitchell claimed that his problem in the dispute was 'to find a mode of action which will be effective.' Given the intransigence of the employers, and the union's refusal to accept employees' committees, this was a difficult task. Evidently compulsory union recognition was still an option that Mitchell and the government rejected. He told the religious leaders that he hoped to arrange a settlement of the dispute over the next weekend. This was a reference to an arbitration proposal that he was about to submit to the parties. Atkinson's sympathy for the Kirkland Lake miners apparently went beyond his mild editorials of support. He gave Reverend Coburn and Reverend John Frank $100 towards expenses incurred in their journey to Ottawa to lobby the government.[148]

Another function of the Board of Evangelism and Social Service was to explain the church's labour policy to its ministers. Coburn and Mutchmor were well equipped to accomplish this task for both men had a thorough grasp of the principles of collective bargaining, the techniques of union organizing, and the history of employer-dominated company unions. They consistently pressed the point that 'the chief responsibility in this case lies with Ottawa and ... we are doing our best to get action there that will be helpful.'[149]

This activity by Mutchmor and Coburn was undertaken because they believed that the Kirkland Lake dispute was part of a national problem about which their church was deeply concerned. As Coburn wrote: 'There may be many issues involved in the Kirkland Lake strike but I am convinced that more important than any of them ... is the issue as to whether labour is to have the right to choose its own agencies for collective bargaining purposes. This is an issue of *national* importance; that is why we look upon it so seriously.'[150] But their concern about labour relations was not only humanitarian; it also was connected to their interest in the war. As in the First World War,[151] the United Church supported a total war effort. By 1941 it was concerned that labour unrest in Kirkland Lake and elsewhere could adversely affect the level of war production. The churchmen believed that the government's failure to solve the labour problem was hurting efforts to support the men in uniform. These church leaders believed that 'there is a colossal and lamentable failure [in production] due to the lack of strong leadership in Ottawa.'[152] For this reason, they supported the idea of strong leadership, a coalition government, and conscription – all of which were policies of the Conservative Party. This may be one reason why their views on labour problems were accorded considerable attention by the government.'[153]

The Conservative Party itself was not particularly sympathetic to trade unions. The federal and provincial Conservative parties supported 'no-strike pledges,' compulsory arbitration of disputes, and wage and price controls. In the course of the Kirkland Lake dispute the Ontario Conservative leader, George Drew, urged Prime Minister Mackenzie King to organize a federal and provincial government conference with employers and organized labour in an effort to end labour disputes that were 'undermining our war effort.' He believed that Canadian workers were loyal and that the right to organize should be assured by law, but he believed equally that the labour relations situation in Canada was 'intolerable' and that the country had a right to prevent interference with the war. He maintained that it was nonsense to speak of an all-out war effort if full production depended on the whim of some 'professional agitator who has no stake in Canada and has no reason to be personally concerned about the serious social consequences of the bitterness he creates.'[154] Like the mine operators, he rejected the notion that the strike was indigenous, and the result of grievances of the local employees. He accepted the employers' argument that the dispute was imposed on the loyal workers by outside agitators. He criticized the federal government for allowing American organizers into the country to create disorder, and then leaving the problem of suppressing the disorder to the provincial governments. Like Mitchell Hepburn, he disapproved of the federal government's policy and approved Hepburn's action of sending in police to quell the workers. In his view, disturbances like these should never be allowed to occur. From the viewpoint of organized labour the Conservative proposals were even more repressive than the Liberal's policy – however inadequate that might be.

The Liberals, of course, did not believe in compulsory arbitration or compulsory collective bargaining, although the two issues were related in their minds.[155] They urged labour and management to support national unity and the war effort, to settle their disputes voluntarily in accordance with the principles of PC 2685, and to resort, if necessary, to the IDI and IDIC machinery. Their approach involved wage and price controls, 'dollar-a-year' business men to run the war effort, 'fair' profit for companies, and accelerated depreciation of new capital equipment used in the war industries. They advocated some social welfare measures but did not set a floor on wages or create a 'low wage' exemption from wage control. Instead they implemented a bonus policy, which initially was applied at the discretion of the conciliation boards and later was applied to all wage earners by the National War Labour Board.

The Communist Party had been banned in Canada in 1931; its continuing legal problems stemmed from its opposition to World War II before the breakdown of the Nazi–Soviet Pact. Front organizations such as the Labour Progressive Party (LPP) had not yet been formed. Nevertheless despite such impediments to organizational activity, the *Tribune*, a communist paper, editorially supported the Kirkland Lake miners' struggle.[156] Locally the communist element in the union remained active and the union's lawyer, J.L. Cohen, continued to be close to the Communist Party.

As might be expected, the CCF organized the most active and consistent political support for the strike. By this time, the CCF labour policy was identical to that proposed by the CCL, and the CCF social policies were widely supported in the industrial union movement. In the summer of 1941 the CCF had helped the Steelworkers' Organizing Committee in its efforts to secure the appointment of a conciliation board for the National Steel Car dispute in Hamilton.[157] In return SWOC officials agreed to assist in distributing a CCF trade union pamphlet at NASCO (National Steel Car Corporation). Shortly thereafter the party became involved in the Kirkland Lake dispute. The caucus urged the federal government to intervene to settle the strike. CCF members of Parliament regularly visited Kirkland Lake to speak to the miners.[158] Clarie Gillis, the member from Cape Breton South, made several visits both before and during the strike. As a former coal miner and member of the UMW, he was able to eloquently express the plight of the strikers to members of the House of Commons and through that forum to the general public. CCF constituency associations and clubs passed resolutions of support for the miners,[159] and made small financial donations to the strike funds. CCF Alderman Stanley Knowles succeeded in getting the Winnipeg City Council to discuss a resolution favouring the recognition of the Kirkland Lake miners' union, which ultimately passed on 4 January 1942. The Winnipeg council insisted that the operators comply with the government's declaration of principles and bargain collectively with the union chosen by their employees. Knowles, by his action, won the 'full backing of the CCL people and the formal support of the TLC.'[160]

Other municipal councils sent resolutions of support to the federal government. The Teck Township Council sent one. On 11 December 1941, the mining municipalities of Northern Ontario unanimously urged the government to take immediate action to settle the Kirkland Lake strike, to pass legislation to facilitate the settlement of similar labour disputes, and to speed up Canada's industrial war effort. They sent a four-man delegation to Ottawa to express their concern over the loss of municipal revenue.

Besides the support of these organized groups, there were occasional demonstrations of public support. In mid January 1942, a meeting of 700 people took place in Massey Hall in Toronto. The six speakers included Reverend J. Coburn of the United Church, Reverend John Frank of the Anglican Church, Reverend Harold Toye of the Christian Labour Forum. The principal labour speakers were Pat Conroy of the Canadian Congress of Labour, William Simpson, president of Local 240, and Forrest Emerson, an international representative of IUMMSW. The meeting contributed $1000 for the miners' strike fund. Many people in Toronto and in other communities wore 'I Support the Kirkland Lake Miners' buttons and held meetings similar to the Toronto gathering. Nevertheless, despite this widespread support for the miners' position, it was apparent that the successful resolution of the strike would require a change in federal government policy.

8
Government intervention in the strike

The reactions of the federal and Ontario governments to the Kirkland Lake strike were not different in substance. Neither was interested in intervening to promote collective bargaining or to protect the rights of the workers to select their own union. The tactics of the two governments, however, were quite different, and reflected the political styles of Mackenzie King and Mitchell Hepburn. Hepburn confronted the situation, while King delayed intervening in the hope that the situation would resolve itself. His government's investigatory machinery postponed the strike for as long as possible, but the only solution advocated by his labour minister was the old industrial council concept, now renamed the 'Kirkland Lake Formula.'

Mitchell Hepburn's reaction was predictable. Since the 1937 Oshawa strike he had been hostile to trade unions. In the war situation, his anti-union feelings were still very evident. Speaking in New York in September 1941, Hepburn lashed out at 'those who cause strikes in war industries' and declared that 'in Canada, there should be no room for such troublemakers. They should be listed, classified, condemned and divorced from society.'[1] He implied that union organizers were fifth columnists and suggested they be tracked down by veterans' organizations that could be mobilized as special police. Several months later Hepburn remarked that he 'was keeping a tight lid on CIO agitation in Kirkland Lake,'[2] and after the strike began, the Ontario government was considered to be 'even more clearly aligned on the side of the owners.'[3] Not one of the government's spokesmen criticized the conduct of the operators, but they condemned the union in unmeasured language. The government sent in a large contingent of police (whom the strikers dubbed 'Hepburn's Hussars'), against the expressed wishes and protests of the local authorities. The police increased tensions in Kirkland Lake, harassed strikers, and assisted management efforts to continue mining operations. There was no doubt concerning Hepburn's sympathies. After the strike, 'both the President of Macassa mines (in whose stocks Premier Hepburn had previously invested) and the Managing Director of Lake Shore Mines thanked Premier Hepburn for his assistance ... in staving off for the present, the taking over of control of operations by the CIO.'[4]

Nevertheless, the primary government role was played by the federal government, which during the war had ultimate legislative responsibility for the regulation of industrial relations. Because of its policies of compulsory conciliation and compulsory strike votes, federal agencies, most notably the IDIC (Industrial Disputes Inquiry Commission), became deeply involved in the Kirkland Lake situation. This intervention delayed but did not prevent the strike, and the federal agencies did not resolve the underlying causes of the dispute. Government policy had traditionally shown 'a marked preoccupation with attempting ... to prevent strikes and avoid "public disorder," rather than with protecting the rights, liberties, and prerogatives of one or the other of the contending parties.'[5] This policy was applied, and failed in Kirkland Lake. As a result the government was called upon to re-evaluate its approach to collective bargaining and to intervene directly in the strike. The government did not in fact intervene, but it did reconsider and ultimately reaffirm its established policy with respect to collective bargaining.

At the time of the Kirkland Lake strike the government was faced with a number of events and political problems that tended to overshadow the situation in the small northern community. Arthur Meighen had assumed the leadership of the Conservative Party, and there was pressure on the government from both the Conservatives and the 'Committee for Total War'[6] to introduce conscription. The Japanese attack on Pearl Harbor resulted in the immediate entry of the United States into the war as a belligerent. The death of Ernest Lapointe was a personal loss to Mackenzie King, and necessitated a reorganization of the Cabinet. Near the end of the strike there were by-elections in South York and Welland. All these events prevented the Kirkland Lake strike from receiving as much of King's personal attention as might otherwise have been the case, for despite his conservatism King was still more interested in industrial relations matters than his Cabinet colleagues. King seemed to realize that organized labour was becoming alienated from his government. He held two full Cabinet discussions on the subject, but then did not follow through with any action. In at least one other instance, other more pressing preoccupations caused the Kirkland Lake situation to be neglected.[7] However, this situation should not disguise the fact that the Liberal government's labour policy had been and remained *opposed* to collective bargaining, or a Canadian Wagner Act. It pursued its objective of minimizing strikes (however unsuccessfully), not by resolving the causes of conflicts or by supporting collective bargaining as a mechanism for their private resolution, but by the tactics of evasion and delay. Business and labour as represented specifically by the CMA (Canadian Manufacturers' Association) and the CCL (Canadian Congress of Labour) held opposing views on national labour policy. The government tried to conciliate these essentially irreconcilable views, and in so doing pleased neither side. The King administration did react to labour's dissatisfaction with the wage policy by proclaiming PC 8253 (see Appendix 1). In contrast to the political success of the plebiscite on conscription, which blunted Conservative attacks, the change in the wage policy was a failure as it involved a wage freeze, and did not, therefore, significantly improve relations with the labour movement.

The government's position prior to the Kirkland Lake strike was carried through until the strike was lost. Its aversion to both compulsory collective bargaining (the labour movement and the CCF position) and compulsory arbitration of all disputes (the position of the business community and the Conservative Party) was outlined on 7 November 1941, just prior to the beginning of the strike, in a highly publicized speech by Norman McLarty, then the minister of labour. These diverse views illustrated a widespread difference of opinion in Canada regarding labour policy and labour legislation. In the absence of consensus, the government was determined to maintain the *status quo*.

McLarty argued that compulsion was not 'the true corrective method or the one that in the final analysis will prove effective in producing stability in the relationship between labour and management.'[8] 'Co-operation' not 'coercion' was the essence of good industrial relations.[9] McLarty maintained that neither he nor the prime minister were against the aims and aspirations of labour (as Angus MacInnis had recently charged in the House): 'I believe that trade unionism is the essence of industrial democracy and that it must be woven ever more firmly into the fabric of industrial relations.' But, McLarty suggested, industrial democracy could not be imposed; it could only exist with employer consent. Nevertheless, while industrial democracy might be a long-term goal, his immediate preoccupation was the elimination of strikes, and he praised 'those labour leaders who by their sincere effort to promote industrial peace' had made the peaceful settlements possible. It was recognized at the time that he was not referring to the leaders of the CCL unions. McLarty admitted that many employers were not genuinely willing to negotiate and sign collective agreements with trade unions' but he had no suggestions as to how, short of strike action, they could be prompted to do so. The government, McLarty continued, did not approve of unions exploiting the war situation 'to "high pressure" workers into a union which without such high pressure methods they would not be willing to join.'[10] This argument was similar to the mine operators' theory as to how their employees were unionized but was, of course, beside the point. There was no difficulty in organizing employees; the problem was to persuade their employers to recognize and bargain with the union they had selected.

The Liberal concept of democratic industrial relations[11] was apparently different from that of the labour movement and the CCF: it was 'undemocratic' to compel an unwilling employer to recognize a trade union. In contrast, the labour movement believed that a union was democratically entitled to represent a group of employees when the majority had voluntarily indicated their support. The Liberal position, like that of the employers in the mining industry, seemed to be an adherence in industrial relations to a policy reflecting a 'unitary framework' of the industrial structure; in a political context, this policy approach was called a belief in consensus. Since the organization of the war effort required a considerable degree of government control, compulsory strike votes and wage controls were an established part of the industrial relations process. Since King himself had been an early advocate of compulsory conciliation, it is doubtful that the government's decision to maintain the *status quo* was really based upon an 'aversion' to compulsion. Rather it did not want to compel *bargaining*, which to the government connoted conflict.

It would appear that the decision to do nothing resulted from a mixture of political and philosophical motivations. The government was concerned to avoid strikes that might prejudice the war effort; and it was unconvinced that the legislation of compulsory recognition would resolve industrial conflict. On the one hand, such a legislated solution was actively supported by the CCF and the labour movement, but it was adamantly opposed by business interests and the Conservatives. On the other hand, the compulsory-arbitration solution favoured by business and the Conservatives had been rejected by labour. These countervailing pressures favoured the maintenance of the *status quo*. In any event, King himself seemed convinced that there was no inherent conflict of interest between employer and employees. This belief in a common purpose and harmony of interest led him to view conflict as unnecessary and due to faulty communications, misunderstandings, outside agitators, or 'irresponsible' leadership. 'National unity,' 'consensus,' and 'industrial harmony' were all key phrases in the Liberal rhetoric, but in the industrial relations sphere at least, they accurately described the beliefs and goals of the prime minister.

The government's attitude was crucial to the outcome of the Kirkland Lake strike, as many realized at the time. J.R. Mutchmor commented that, 'The present attitude of Ottawa is "wait and see" ... We must continue to do our utmost to break down this lethargy. The rising tide of public demand for strong federal leadership may compel the Prime Minister to retire from the scene. It is certainly true that nothing of any constructive value will happen in Kirkland Lake unless we can get Ottawa to move.'[12] Mutchmor was a supporter of strong war leadership and conscription and may have overstated the case; however, his reading of the situation was accurate and was one shared by both the strike leaders and the leaders of the CCL. Prior to and throughout the strike the federal government was urged to intervene to *compel* adherence to the principles espoused in PC 2685 endorsing union organization and recognition. Indeed, on 22 October, just after the CCL had officially decided to support the impending strike, a delegation of trade unionists met with the minister of labour to explain the necessity of making the principles of PC 2685 mandatory. The minister promised only to place the views of the congress before the government. Nothing happened as a result of this meeting.

As the strike began, M.J. Coldwell, Angus MacInnis, and Clarie Gillis of the CCF caucus asked the government to clarify its position. Coldwell and MacInnis wrote to Mackenzie King and McLarty pointing out that the Kirkland Lake strike was assuming 'serious proportions,' which if prolonged would be disastrous to labour relations in Canada and, indirectly, to the war effort. In their view the only issue involved was 'the refusal of the mine operators to follow your government's policy and to accept the unanimous recommendation of a Board of Conciliation and the democratically expressed will of the majority of their employees.' Union organization and collective bargaining were accepted instruments in Britain and the USA, and the prime minister himself had declared them to be fundamental to democratic institutions, yet the mine operators were being permitted to 'flout the expressed wish of your government and to jeopardize national unity by fomenting class strife.' Both writers referred to Ontario's recent police intervention and their belief that it was potentially provocative. They urged King to intervene personally just as President Roosevelt had done when a dispute in the coal mines had threatened American defence production:

The dispute in Kirkland Lake presents you ... with a great opportunity to win the confidence and regard of organized labour here and everywhere else, if through your personal intervention the miners of Kirkland Lake could win collective bargaining –

through the union of their choice, then you would win from the democratic world the same measure of admiration and confidence which Mr. Roosevelt has won through his action in the captive coal dispute.[13]

This appeal was meant to flatter King by comparing him to his friend, the American president, whom he greatly admired. What is interesting is that had their advice been taken, it might seriously have affected the good relations between the CCF and the labour movement. Their assessment of the importance of the Kirkland Lake dispute to the government's future labour policy and to the relations between the Liberals and organized labour was more accurate than Prime Minister King's turned out to be. They had come to understand the aspirations of the labour movement 'whose views we have taken particular pains to ascertain.' Unlike the Liberal Party, the CCF[14] had been deliberately involving itself in labour's organizing activities and its legislative problems since 1937. The three CCF members indicated that the settlement of the Kirkland Lake dispute could restore the confidence and trust of the labour movement, but they warned the prime minister that it was essential to 'act before it is too late.'

On 4 December, McLarty responded to Coldwell's letter. He was both unsympathetic and blunt, suggesting that the union had misled employees by suggesting that through collective bargaining they could achieve wage increases beyond those legally permitted: 'There can be no doubt that the union's success in organizing employees was in large measure due to the hope of higher wage rates, which under the order in council were precluded.'[15] This interpretation not only oversimplified the reasons why employees had joined the union, but also ignored the fact that, by this time, recognition had become the issue. There had not been any negotiations concerning wages. McLarty claimed ambiguously that 'the statement that the operators are flouting government policy is open to question,' for 'it has been pointed out on a number of occasions that this order is advisory and not mandatory.' Indeed, suggested McLarty, unions had occasionally engaged in unlawful strikes and a reconsideration of its collective-bargaining policy might entail a prohibition on strikes:

The government has pursued this policy of non-compulsion during the war period and has tried to deal out even justice to both parties. It is felt that should employers be compelled to recognize and deal with a trade union that the question immediately arises as to whether employees should be forbidden to go on strike. If compulsion is adopted it must be applied to both parties and where rights are given, responsibilities should definitely exist.[16]

This nonsequitur reflected either McLarty's personal view of the political trade-off necessary to balance labour and business interests, or his failure to understand the bargaining process. What makes collective bargaining work is the *threat* of strike action. To grant collective-bargaining rights and then insist that anything not resolved by this process must go to arbitration would, of course, have emasculated the process of bargaining.[17]

Mackenzie King's reply to Coldwell and MacInnis was more diplomatic, although in substance it indicated the same reluctance to intervene. He had found it difficult, because of other events, to take on 'more in the way of personal responsibility,' and was not convinced that his personal intervention would help: 'Had I believed that intervention by myself, personally would have helped to improve matters rather than to further complicate the situation, as I believe would have been the case, I would notwithstanding and despite possible consequences have prepared to take this step.'[18] He never elaborated about what he thought the 'possible consequences' might be. Certainly, by this date, there was nothing 'complicated' about the dispute. The only issue was the employee's desire for recognition and collective bargaining through the union of their choice,[19] so much so that the strike had clearly become a test case for that right. He made it clear that he would only intervene when *both* employers and employees welcomed such a course: 'Without such mutual agreement on personal intervention, I doubt if aught of good could be accomplished.'[20] Without this minimum degree of consensus, King believed that his intervention would have been fruitless. This assessment may have been correct, but it is also possible that the intervention of the prime minister might have induced the employers to negotiate. However, King was unwilling to act in so polarized a conflict, particularly after the failure of the conciliation machinery. Moreover, as has already been pointed out, he may well have regarded the 'Kirkland Lake Formula' as an appropriate compromise.

LABOUR APPEALS TO GOVERNMENT

On 29 November, the government received a telegram from the Kirkland Lake conference, which the labour movement had organized in an effort to mobilize national support for the Kirkland Lake miners. On 1 December Prime Minister King's secretary, Mr W. Turnbull, informed King of this telegram and the conference request for a meeting with the prime minister.[21] The next day the Cabinet discussed the Kirkland Lake situation, the conciliation board report, and PC 2685. The minister of labour (and his advisors) in the privacy of council admitted that the companies were in the wrong. King recorded in his diary that the majority agreed with this view yet 'it was impossible to get the Cabinet to decide anything.' After the meeting, King left correspondence concerning the situation in McLarty's hands and suggested he try to work out some solution. Nothing happened. On 5 December Prime Minister King spent a busy day interviewing Louis St Laurent as a replacement to the late Ernest Lapointe. That evening when he turned once again to what had by that time become the most pressing labour problem of the day, he complained in his diary that he was 'too tired to tackle [the] Kirkland Lake mines situation.'[22]

On 6 December 1941, King finally re-drafted his wire to the miners in response to those that had been sent after the Kirkland Lake conference. He consulted McLarty, who approved the final wording. He told the local that he had discussed with the minister of labour and his other colleagues 'whether intervention on my part in the existing dispute at Kirkland Lake could be of service to the parties to the dispute.' He then stated his position from which he never subsequently deviated: 'In view of the steps already taken by the Department of Labour to avoid the strike which has arisen and to effect a settlement by conciliatory means, I have ... come to the conclusion that without mutual agreement by the parties to the dispute, *further intervention by the government could be of little avail at the present time by way of settling existing differences.*'[23]

The labour movement, in backing the Kirkland Lake strike, had taken a calculated risk. It knew that government intervention was crucial. If the government maintained the position taken in its telegram, the strike was lost. Management knew that if it held out without recognizing the union, and the government did not enforce PC 2685, they would eventually win. The only recourse for supporters of the strike was to try to persuade the government to change its position, and this was the objective of lobbying activities undertaken by labour, church groups, and the public over the next two months. On 10 December, J.R. Mutchmor cancelled a trip to Kirkland Lake in order that he might have more time to 'concentrate on Ottawa.'[24]

The government's decision to do nothing was not taken without thought. The deteriorating situation in Kirkland Lake prompted a reevaluation of the government's existing policy and the apparent alternatives. On 3 December 1941, McLarty outlined the government's choices for the prime minister in an extremely important memo. It is worth examining this memo in some detail because it explains why Wagner Act principles were not adopted in 1941 despite considerable public pressure. King's action on 6 December suggests that the prime minister had read this memo and was following his minister's advice.[25]

McLarty began by stating that the government had already done everything 'possible' to avert the strike in Kirkland Lake. He reviewed the work of conciliation officer Campbell, the IDIC, and the conciliation board, and admitted that the mine owners' withdrawal from the board's proceedings had been 'inopportune and unwise' (although he blamed the companies' lawyer for this precipitous action). He concluded that it was this withdrawal, rather than the degree of union support, which had resulted in a unanimous conciliation board report recommending union recognition. Although the union had 'won' the government-supervised strike vote, McLarty cautioned that 'in fairness it might be pointed out that 1,700 men had enlisted in the armed forces in Canada. There was no practical method by which the vote of these men could be registered.'[26] There was, of course, no reason for them to vote, as they were no longer employed in the mines. The comment may indicate that McLarty may have believed the companies' assertion that the union was illegitimate in part because it was dominated by unpatriotic 'foreign' elements.

McLarty chronicled the breakdown of the final mediation efforts prior to the strike and concluded that 'the effective cause of the strike was the question as to whether the union should be in a position to call in outside representatives in their negotiations with various companies.' This was a generous view of the mine owners' refusal to recognize their employees' union, and suggests that McLarty subscribed, at least in part, to the 'outside agitator' theory of the conflict. In any event, McLarty was convinced that, despite the conciliation board's conclusion that the existing machinery was inadequate to resolve recognition disputes, 'the government has sought every method of peaceful settlement of the dispute.' He then proceeded to outline his view of the policy options then open to the government.

The first course was to compel the eight companies in Kirkland Lake to recognize the IUMMSW (Mine Mill) as the sole bargaining agent for the miners, a policy that he believed would be popular with 'certain branches of labour but I doubt if it would be with all.' McLarty was being optimistic. The CCL unions desperately wanted this solution, but by 1941 the TLC (Trades and Labour Congress) also favoured it. McLarty recognized that, in effect, this option would involve acceptance of the principles of the United States Wagner Act, and in his view 'that Act has not been entirely successful, and has resulted in stirring up a large number of jurisdictional feuds.' This was not an accurate evaluation. There were jurisdictional disputes in the United States because the CIO had broken away from the AFL (American Federation of Labor) and had sought to organize the mass-production industries on an industrial union basis. When the CIO experienced some success, the AFL was forced, belatedly, to abandon to some extent its restrictive approach to organizing and broaden its base among unskilled workers. This brought it into conflict with the CIO, since the two central organizations both had chartered unions which were competing for the same members.

Policy differences exacerbated this jurisdictional conflict, but essentially it was the result of structural changes within the labour movement rather than legislative changes. The Wagner Act, once it was declared constitutional by the American Supreme Court, had virtually eliminated recognition strikes. In that sense, it was *very* successful in eliminating situations like the one in Kirkland Lake.

McLarty was opposed to introducing compulsory collective bargaining, and that was the advice he gave to Mackenzie King. Such a solution could not apply only in Kirkland Lake but would have to 'form part of a much wider policy and be applicable to all industries.' Accordingly it would have broad ramifications, since CCL organizing was continuing apace. He also added that if employers were to be compelled to recognize particular unions, they must be given information about them. Unions should be compelled to register the names of their officials, the composition of the membership, a financial statement of funds on hand and salaries paid to union employees, and the names of organizers. There should be assurance to business that a union would assume financial responsibility for any breach of any contract to which it was a party. 'If compulsion must be resorted to, it should be done equally on both sides.' In expressing these views, McLarty was reflecting a management concern of the period about the new industrial unions. These unions were not well known to the government and like the businessmen, the government was suspicious of their stability and their 'responsibility.' This concern on the part of management was sometimes genuine, but often was merely an effort to place unions in a position where they could be subject to civil suit – for at common law a trade union was still a civil conspiracy in restraint of trade.[27] Trade unionists responded that they were open organizations whose books could be examined at any time. They preferred that collective agreements be enforced by the grievance/arbitration procedures incorporated in the contracts, rather than through the civil courts.[28]

McLarty repeated the view he had earlier expressed that compulsory collective bargaining would have to be accompanied by 'a correlative measure of compulsory arbitration and the outlawing of all strikes.' McLarty may have been thinking of the no-strike pledge by the American labour movement as its contribution to the war effort. However, the situation of the two countries was quite different. In the United States, the no-strike pledge had been taken by trade unions long *after* the passage of the Wagner Act, not as a result of it. This pledge was the result of the United States entering the war as a belligerent in late 1941. It was not a policy imposed by the American government on the labour movement, but was agreed to by organized labour after considerable consultation as an equal partner. The CIO movement in the United States was part of Roosevelt's New Deal coalition. Labour participation on war policy-making bodies was significant,[29] unlike that of the Canadian labour movement. American unions had already achieved the status and legislative support that organized labour in Canada now sought.

Apart from this American situation, it is not clear why McLarty connected compulsory collective bargaining and compulsory arbitration of disputes. It appears that he feared that the introduction of Wagner Act principles would inevitably lead to more industrial conflict because it would facilitate trade union organization (an unpalatable result, apparently). This assumption did not necessarily follow. The Wagner Act eliminated recognition strikes. Similarly, in 1944, after the proclamation of PC 1003, Canada's collective-bargaining legislation, strike activity actually decreased. Perhaps McLarty simply wanted to balance political interests and had not thought through the recommendation in terms of its impact on industrial relations. What is clear is that he and his government colleagues did not view collective bargaining as a positive good for society or the economy. Their concern in industrial relations was solely to prevent strikes. The United States had a different tradition. The Wagner Act had been passed in the depression years as a mechanism to help stabilize the economy, raise the status of employees, and create a counterbalance to the power of business. Collective bargaining was, therefore, seen as 'a good thing' for the economy and a democratic society. If, by balancing the influence of different interest groups in society, it thereby reduced conflict, that was simply a beneficial secondary effect.

McLarty knew that the introduction of a policy of compulsory collective bargaining would be relatively simple: 'In short, it would require the substitution of the word "shall" for the word "should" in order-in-council PC 2685.'[30] King, of course, accepted McLarty's advice and this simple procedure was never implemented.

The second possible course of action open to the government in the Kirkland Lake dispute was to appoint a controller for the gold mines. McLarty warned that the CCF would approve of this step 'because it is a definite step in the direction of state ownership and control of industry.' In fact, as a solution to the Kirkland Lake strike, the CCF preferred the introduction of compulsory collective bargaining. The CCF had suggested a controller, but only when it felt it might be an *indirect* way of achieving a change in the government's labour policy, or of forcing the operators to recognize the union.

McLarty believed that the appointment of a controller would eventually lead to a government take-over of every important mine in the country. He suggested that this course would not necessarily settle the dispute but would 'probably result in greater difficulties.' He cited the National Steel Car dispute as an example where 'even after taking over this industry the question of recognition of the particular union was not solved.' This was a telling point, for (as at National Steel Car) a government controller might be put in the embarrassing position of ignoring PC 2685.

The third alternative McLarty presented to Prime Minister King was to let the strike run its normal course. He claimed that he did not mean by this that the government should complacently do nothing, but that is what it amounted to. 'I do believe that we should continue to press upon the operators the acceptance of the formula which would enable Messrs. Mosher, Conroy, McLean to sit in at any conference with the men's representatives.' This formula by-passed Tom McGuire and the local's executive, and substituted the CCL leaders as union representatives of the striking miners. It had already been rejected by the mine operators on 18 November.[31] If there was further government pressure on the mine operators to accept this formula, there is no evidence of it, and such negotiations never came about. McLarty believed that 'the danger' of proceeding with this procedure was that it might become a precedent for other organizations that 'are members of the CIO.' He believed the most likely areas for such 'difficulties' were the steel industry at Algoma and Sydney, the auto industry at Oshawa and Windsor, and the coal-mining industry in Nova Scotia and Alberta. To present the problem in this manner to King, who instinctively procrastinated over all political controversies, undoubtedly made complete inaction seem more attractive. Finally, McLarty told King that he had been advised that day that 'men are returning to work in increasing numbers in Kirkland Lake and that the strike will probably be over in a very short time.' This, he cautioned, was 'ex parte information' (i.e. probably heard from the management side), and turned out to be incorrect. The strike dragged on for two more months. Nevertheless, it was one more reason for the government not to intervene.

McLarty concluded his memo with the opinion that 'there is not sufficient time to enable us to adopt a well-rounded policy, such as that envisaged in the suggestions of Part I.'[32] The policy in Part II was 'inadvisable.' The government was, in his view, forced to fall back on Part III – inaction.

Hence, by early December, the government had *firmly* decided not to intervene in the Kirkland Lake dispute and not to change its policy on collective bargaining. Only once after the strike began did it make an effort to settle the strike. Otherwise nothing was said or done. This one effort was an attempt by the new labour minister, Humphrey Mitchell, to end the dispute by voluntary binding arbitration.

The labour movement, of course, had no idea how firmly the government's policy on non-intervention had become. It had no alternative but to press for action. Consequently, the Kirkland Lake conference was followed by a conference in Ottawa that, on 13 December, established a national Kirkland Lake committee. King had not replied to the statement emanating from the Kirkland Lake conference or to the delegates' request for a meeting with him. On 7 December, therefore (between the Kirkland Lake and Ottawa conferences), Local 240's President William Simpson sent another telegram to the prime minister. He pointed out that despite King's lack of response, the union was hopeful that the prime minister would meet and listen to their submission.

International events then intervened. The Japanese bombed Pearl Harbor and the Americans entered the war. The Ottawa conference on the Kirkland Lake strike took place, but King still did not respond to the request for a meeting. On 15 December the unions, for a *third* time, requested a meeting with Prime Minister King: 'The anti-labour policy of the Kirkland Lake mine operators and the government's toleration of that policy, have seriously disturbed the workers ... and undoubtedly affected their morale, at a time when it is essential that the strongest possible basis for the fullest measure of popular support of every phase of the nation's war effort and war purpose be established.'[33] No direct reply was received from the prime minister, but information was obtained from his office that he had requested his new labour minister, Humphrey Mitchell, to see what could be done to obtain a settlement.[34]

On 30 December 1941, William Simpson again requested that Prime Minister King intervene in the strike and the CCL reiterated its request for a conference. By 8 January 1942, there still had been no reply. The labour movement by this time concluded that the Kirkland Lake miners were striking against impossible odds. They were picketing the mines in temperatures below zero; they lacked sufficient food and fuel; they were up against influential anti-labour newspapers and 'the Government of Canada as well, since the Government does not seem to be interested in enforcing its own declared policy on collective bargaining.'[35]

Finally, when the prime minister did reply his message was similar in content and tone to that of 6 December, except that now he was more definite. On 13 January 1942, he advised that unless Mitchell's efforts to settle the strike were successful, 'no useful purpose would, so far as I am able to see, be served by intervention by myself or other members of the government.'

Developments since his earlier telegram of 6 December had strengthened his belief that his intervention would be of no use: he concluded that 'attempts at further intervention might indeed only tend further to prolong the dispute and to delay its ultimate settlement.'[36] This was, of course, true in one sense. The absence of government intervention shortened the strike, because it would soon be lost, but it 'settled' neither the dispute nor the general issue of recognition, which consequently continued to reappear in later strikes.

Pat Conroy (CCL secretary-treasurer) expressed his regret that the prime minister was 'apparently unwilling' to discuss the application of the government's labour policy in Kirkland Lake. The principles of PC 2685 were not respected in either the war industries operated and controlled by the government or in private industry, and it appeared that the Canadian government would do nothing. Conroy requested a meeting to discuss not only the Kirkland Lake situation, but also the government's labour policy as it affected the general issue of union recognition and collective bargaining.[37]

In the midst of this correspondence an event of great significance took place. Norman McLarty was shifted from the labour portfolio. King was concerned about the failing health and age of his minister of labour. He also wished his labour minister to better represent the voice of organized labour in the Cabinet. On 15 October 1941, King promised labour leaders that the next labour minister would be a trade unionist, and he had, in fact, begun to search for a replacement for McLarty during 1940.[38] As labour's discontent grew and the labour movement became increasingly critical of the labour minister himself, the need for a new man became more urgent. King had informally approached TLC president Tom Moore on several occasions and had offered him the job. Moore was a conservative business unionist and had a good relationship with the prime minister. The TLC prior to the loss of the Kirkland Lake strike was less militant on the question of compulsory collective-bargaining legislation, although after the Kirkland Lake strike its position hardened and became the same as that of the CCL. On 11 December 1941, Prime Minister King formally offered the position to Moore: 'Moore refused politely, stating he had to keep the labour movement together, an especially important task in view of labour's feeling that the government was not playing fair with them.'[39] He suggested that King approach Humphrey Mitchell. The next day King told McLarty that 'he had not done himself full justice' in the labour portfolio and that King had decided to make him secretary of state, that C.D. Howe (minister of munitions and supply) preferred Mitchell to Moore and that, in any event, Moore had rejected the offer. McLarty agreed that Mitchell was a good choice. He was known to have a working-class background and trade union experience, although he had few admirers in the CCL.[40] Mitchell accepted the job and on 5 December 1941, he was sworn in as minister of labour. King liked Mitchell, felt correctly that their views on labour relations were close, and expected he would do well in the post. On 16 December 1941, King announced his new appointment. President Moore (TLC) publicly said he was pleased. President Mosher of the CCL was ambivalent.[41]

A NEW MINISTER OF LABOUR

The appointment of Humphrey Mitchell as the minister of labour was a disastrous blow to the Kirkland Lake miners. Although he was a man trusted and admired by King, Mitchell's credibility with organized labour was seriously undermined by his role in the Kirkland Lake dispute as head of the IDIC. Mackenzie King apparently remained unaware of the CCL's deep distrust of Mitchell, perhaps because he had not established contact with the CCL as he had with the TLC and he did not like the new union leaders because they were too aggressive. He wrote of Mitchell's appointment: 'It will afford the kind of link with labour which for so long I have felt in so great need.'[42] After Mitchell became the minister of labour, Local 240 could expect little assistance. For Mitchell to actively intervene in the dispute would have meant public repudiation of his earlier settlement proposals made when he was chairman of the IDIC. His actions throughout the strike remained entirely consistent with his earlier position.

The Kirkland Lake situation, with a new minister of labour at the helm (as yet not elected to the House), was once again discussed in Cabinet on 16 December. King raised the issue and urged the Cabinet to back Mitchell in an effort to get an immediate settlement. He suggested that Howe and Crerar (minister of mines) tell the operators that unless they assisted the government in reaching a compromise agreement, they would be faced with the settlement of the dispute by compulsory arbitration. Apparently King preferred to use compulsory arbitration rather than compulsory collective bargaining as a threatened solution – perhaps because arbitration could be more easily confined to this one situation. At the same time he instructed Mitchell to persuade the labour leaders to compromise. King spoke frankly to the Cabinet, telling them that the owners were wrong and that the government was wrong in not taking 'sterner' measures to support their own policies and legislation. Mitchell spoke 'very sensibly' about the situation and began taking matters in hand. King put great faith in Mitchell; their approach to industrial relations was similar. It was a matter of cajoling the parties, of finding the right formula, of getting the parties to compromise their differences. On 22 December King wrote to a constituent that he was 'hopeful that Mr. Mitchell is going to find it possible to bring about a settlement to the dispute.'[43]

After becoming minister of labour, Mitchell periodically held discussions with representatives of the two sides in an attempt to search for a basis of compromise. He began to probe the union about the possibility of settling the dispute by arbitration.

The first time that the question of arbitrating the dispute had arisen was at the beginning of the strike when the Teck Township Council passed a resolution asking the federal government to investigate the possibility of setting up an arbitration board. At the council meeting, the union's lawyer, J.L. Cohen, said that the union would be willing to abide by the decision of an arbitration board. The operators made no comment on the proposal. Following this meeting, Local 240's president, William Simpson, wired Prime Minister King and McLarty that the union agreed with Teck Township's arbitration proposal: 'I wish to assure you that this union is ready now, as it always has been, to settle this dispute peacefully and if a *properly constituted* board is set up we are prepared to co-operate with and abide by the decision of the board.'[44]

On 28 November, the operators rejected the arbitration suggestion of the municipal council. Nothing further occurred until 9 January when Mitchell proposed a formula which, in his opinion, would 'dispose of the unfortunate issue which has developed at Kirkland Lake' and which would also accommodate the principles of PC 2685. This formula was a variation of that suggested by the IDIC and by McLarty just before the strike began. The men would return to work without fear of discrimination. Within thirty days they would elect five 'representative employees' to negotiate an agreement with *each* mine to govern wages and working conditions. The election would be supervised by the labour department. Each elected committee could be assisted in negotiations by a member of their organization, but the delegation could not include more than two 'outside persons.' The operators were to have the same privilege. Any disputes arising concerning the interpretation of the negotiated agreement were to be referred to the National War Labour Board (NWLB) whose decision would be final and binding. If points arose during the negotiation on which there was no agreement these too would be referred to the NWLB. The NWLB had been set up to administer the government's wage policy and, by this time, included labour representation. In a later proposal, it was suggested that a tripartite board of arbitration perform the role originally envisaged by the NWLB. In a still later proposal the employees would return to work once the board of arbitration was set up and not before. These changes had the effect of emphasizing the arbitration of the dispute rather than any prior talks between the representatives of the parties. In all likelihood, none of these proposals would have been acceptable to management.[45]

Simpson replied to Mitchell's arbitration proposal. The union had previously been willing to submit issues in the strike to impartial arbitration, and remained willing to consider any other proposal consistent with the basic right of labour to bargain collectively through representatives of its own choosing. But the proposal as submitted still gave the operators, not the workers, 'the controlling influence in selecting [the] collective bargaining medium of employees.' The proposed bargaining structure was unclear, but seemed to imply employee committees that would negotiate separate agreements with each mine. The union remained prepared to enter discussions with the operators or the government to conclude some arrangement that was consistent with the public interest and public policy enunciated in government order-in-council PC 2685. President Simpson concluded dramatically:

Our strike, which has received the unqualified support of the Canadian labour movement and all public minded citizens, is in defense of basic principles essential to the well-being of the Canadian nation and its war effort. We are prepared, as we have been since the outset, to compromise on mere matters of form or questions of personal interest but no formula could in honour be accepted by us which violates the fundamental principles of free collective bargaining. We are confident of victory but even if faced with defeat we could not adopt any course which betrays the confidence and support which the Canadian people have so generously extended to us.[46]

FINAL UNSATISFACTORY NEGOTIATIONS

Public pressure in support of the Kirkland Lake miners continued to develop. A *Toronto Star* editorial of 14 January 1942 attacked the view that the CIO was an irresponsible organization. Its affiliated unions in Canada had a large membership. It had a number of agreements with employers and was negotiating at that time with the Ford Motor Company, one of the biggest companies in Canada. In the United States, leaders of the CIO were on government committees, making decisions on the conduct of the war: 'Thus some of the greatest industries in the United States sit down with the CIO officials. And that is not an isolated case.' The United States' National War Labor Board had its labour representation equally divided between the AFL and the CIO:

The President of the United States recognizes both, as accredited representatives of labour. The attempt in some quarters in Canada to give the impression that the CIO is an organization not to be dealt with ignores the fact that employers are dealing with it all over this continent, and that CIO representatives are sitting upon official defence bodies in the United States, together with the foremost representatives of U.S. industry.[47]

Public pressure from organized labour, the churches, the CCF, and occasionally the media prompted the government to propose another arbitration formula even though there was little likelihood of the mine operators agreeing to arbitration. The mine operators could achieve their objective of defeating the union by simply refusing to meet until the strike crumbled. They stuck to their position and the government could not move them.

On 16 January, CCL President A.R. Mosher was requested by the minister of labour to see if the miners would agree to call off the strike on the condition that the dispute would be referred to the NWLB. The men would, in those circumstances, be reinstated in their former positions in order of seniority, without discrimination. The NWLB decision would be final and binding. Mosher communicated this proposal to the local, which consulted with J.L. Cohen. Reid Robinson's position was that 'we had previously offered to put all points of controversy to arbitration.' The union was confident of the merits of its case; moreover it recognized that its position was weakening.

The local union was, at first, reluctant to accept arbitration; they were persuaded to do so only after the international union executive unanimously recommended acceptance. On 18 January, Simpson informed Cohen that the local had accepted Mitchell's proposal at a special membership meeting held that day. Cohen informed the minister of labour.

In the meantime, on 19 January, Mitchell had met with the mine operators to receive a reply to his proposal of a settlement by arbitration. The operators told him they wanted three days to formulate a reply. On 19 January, newspapers across the country reported that the minister of labour had suggested that the NWLB arbitrate the dispute, and that the union had agreed. Management had not. The next day, the same papers quoted the mine managers as denying that any proposal had ever been made to them. This action set off a flurry of activity. Mosher and Cohen both contacted Mitchell who was annoyed by the mine operators' statement, but thought that it was merely an effort to offset the publicity given to the union's acceptance of the minister's arbitration proposal the previous day. Cohen suggested that the entire matter would be cleared up if Mitchell issued a statement to the press to the effect that he was expecting a reply from the operators in a day or two. Mitchell ultimately decided to phone the committee of mine owners the next morning (21 January). Mitchell was optimistic about getting the dispute to arbitration. Cohen was less so. Nevertheless, on 12 January, the minister's statement appeared in the press, making it clear that the arbitration proposal had been submitted to the owners and to the union.[48]

It was apparent that the mine operators were considering the arbitration proposal, and Mosher was unwilling to press for the operators' reply for fear it would show too much eagerness. J.L. Cohen, however, saw no reason for delay. Accordingly, Simpson, on behalf of Local 240, wrote to Mitchell reminding him that he had promised that the mine operators would meet and inform him of their reply. A week passed and the minister had still not advised whether the mine owners had accepted or rejected the offer.

On 23 January, Cohen called Mosher who was outraged by a letter written by Mitchell on 21 January (i.e. after he had met the mine operators) directly to Local 240's President Simpson to 'confirm' the points of his arbitration proposal. They were similar to those earlier presented to Mosher *except* for one very important addition. Once the mines were operating normally again, 'the matter of *methods to be adopted to establish negotiation relationships* between the miners and each of the mines respectively by which they are employed, for the purpose of concluding agreements to govern wages and working conditions of the miners will be referred to the NWLB.'[49] In this proposal the board would not simply decide the terms of settlement (which is what the union had agreed to), but would clearly have the discretion to decide the *kind* of negotiating relationship that would exist. The earlier arbitration proposal had allowed for a union role in negotiations – albeit on a mine-by-mine basis. This proposal contained no such guarantee. The union feared that the earlier IDIC proposals of *independent* employees' committees might be recommended by the NWLB (of which Mitchell was chairman) as fulfilling the principles of PC 2685. Such a recommendation would be binding. The local would have agreed to its own liquidation.

Cohen agreed with Mosher that the contents of the letter were 'in complete variance from the original proposal.' There followed 'quite a hot exchange' with Mitchell as to whether the 21 January letter was in accordance with the original proposal. On 24 January 1942, Mosher followed up this discussion with Mitchell with a letter, in which he said, 'I should like to point out to you, as I did in my conversation with you yesterday, that I regard the contents of that letter as embodying a proposal which is at variance with that which you requested me to submit to the miners and which I informed you subsequently had been accepted by them.' He described the new proposal as 'clearly and obviously a wholly different proposal from the one which at your request I submitted to the miners.'[50] The first proposal put *the entire dispute* before the NWLB for final disposition. The new proposal provided that *only* the method of carrying on negotiations would be dealt with by the board, and indicated that the NWLB would have considerable discretion concerning that issue. Mosher protested that he should have been consulted about the change: 'Your action in submitting a *new* proposal directly to the miners was most objectionable from the standpoint of both courtesy and ethics.' Mosher felt that Mitchell had taken advantage of his relationship with the miner's union,[51] and he was concerned about the public reaction. He determined to make the facts known in order to clarify his role in the situation.

After talking to the mine operators on 21 January, Mitchell had apparently decided to modify his arbitration proposal so that *if* the union accepted it, there would be a greater chance of the operators' acceptance.[52] His letter to Simpson was dishonest. His conduct with Mosher was inconsiderate. His second proposal, however, was predictable. As with the IDIC proposal and his first suggestions of arbitration earlier in the month, he evaded the question of union recognition and tried to gain the acceptance of alternative proposals on which the parties could compromise. He wanted to settle the strike without facing the inherent problems causing it. The NWLB could reintroduce the Kirkland Lake Formula, which Mitchell approved and which he knew the operators would accept.

At the time of this development, the government was immersed in the question of the plebiscite on conscription recently announced in Parliament. Because of an emergency conference of the Cabinet, Mitchell was unavailable and unable to meet with CIO President Philip Murray, who together with J.L. Cohen and union representatives was seeking an interview. Murray's intervention was one more attempt to exert pressure on the government. Philip Murray, the CCL, and J.L. Cohen all had tried (until January 26) to reach Mitchell, but without success. They made it clear, however, that Local 240 had agreed to Mitchell's original arbitration proposal, but not to the later, modified one. The CCF caucus raised the question of the mine operators' position on the arbitration proposal in the House, but the mine operators had made no formal reply to either proposal.

On 23 January 1942, councillors C.C. 'Doc' Ames and Tommy Church secured passage of a resolution in the Teck Township Council that called on their local MP and the government to adopt a policy 'which would give labour its recognized place co-operating with industry for a total war effort.' Ames pointed out that the government had proposed arbitration and while the union had accepted, the mine operators had not yet indicated their stand.

The operators never did take a stand publicly. Privately they had refused the government's arbitration proposal. This was the government's last effort to settle the strike. As the strike wore on there were persistent rumours that the world market for gold was declining and that, following the entry of the USA into the war, the industry would decline in importance.

On 2 January 1942, the Kirkland Lake miners sent a petition to the prime minister asking the government to clarify the status of the gold-mining industry. In the event that the government decided gold mining was not an essential war industry, the strikers asked to be transferred to other essential war work, preferably in the base-metal mines. This petition was signed by more than 2000 miners[53] who were worried by the mine operators' warnings during the strike that they would not be rehiring after the strike was ended because of reduced production. The strength of the union's bargaining position was largely dependent on the role of gold mining in the war effort. The industry's changing fortunes could not have been foreseen before the strike, but had a bearing on the government's refusal to intervene and, therefore, on the local's decision to end the strike.

The government promised to give careful consideration to the miners' petition. The prime minister advised William Simpson that the problem of labour transference could not be considered in relation to one particular industry, but only in relation to a general manpower policy of the government 'to which urgent consideration is now being given.'[54] The government's response indicated that nothing would occur before the end of the strike. As it happened, even after the strike the matter was left to the free play of the market. The hardships of the unemployed in Kirkland Lake were not relieved.

Prime Minister King discussed the Kirkland Lake situation with the leaders of the CCL for the last time at the end of January 1942. On 28 January, King was informed that Mosher and a committee of three or four from Kirkland Lake were in Ottawa seeking a meeting: 'Mr. Mosher and others of his organization have been put off on many occasions with the words that Mr. Mitchell was hoping to make some progress in the settlement, but they did not seem to think they were getting the attention they deserve.'[55] On 31 January 1942, the prime minister agreed to see the committee. The meeting actually took place on 4 February. The national Kirkland Lake committee had met with Mitchell the previous week.

Two interpretations of this meeting have emerged. Mackenzie King himself inserted a formal two-page memorandum in his diary giving his account of the meeting. He recounted how, on 3 February, MPs Coldwell and MacInnis had asked him to meet Pat Conroy and Reid Robinson to discuss the Kirkland Lake situation. At first he refused, saying he had always been willing to 'lend what good offices' he could to avoid the strike, but he believed his intervention would worsen rather than improve the situation. Further, he did not want to appear to be going over the head of his labour minister. However, he was advised that 'it was not with a view to having any intervention; that the men knew they were defeated; that the strike was practically at an end, and that all they were anxious to see me about was for the Government to use what good offices it had to see that the men would not be victimized as they began to return to work.'[56] Accordingly, King agreed to see the delegation, but warned that the most he could do was listen. He 'could not say they [his good offices] would be in the least effective as [he] did not know what attitude the mine owners might take.' On a slightly bitter note, King told Coldwell that 'some of the men involved in these mines were even more bitter towards myself than they were towards the mine owners, and that I did not see that any request by me was likely to be of much help though I would be quite prepared to listen to a statement along the lines they had indicated.'

The meeting took place the next morning with his secretary, Mr Turnbull, present. King began by telling the delegation that he had agreed to meet them at the request of Messrs Coldwell and MacInnis 'with the understanding that it was not to intervene in any way, but simply to listen to what they might wish to say to me in the absence of the Minister.' According to King's account, both Robinson and Conroy told him that they were anxious to assist in the war effort, but that unless the Kirkland Lake strike was settled, there was danger of other mine workers across the country going on strike and thereby injuring the war production.

King listened first to Robinson, then to Conroy, apparently saying nothing. After they had concluded their remarks, he repeated the terms on which he had agreed to meet them – 'on the distinct understanding that there would be nothing said that would relate to any further intervention in the matter of bringing the strike to a close.'[57] He then told Robinson and Conroy that as the prime minister he could not countenance *threats* with respect to sympathetic strikes or other illegal activity. He concluded that he could say nothing more and had other matters to attend to. He shook hands with both of the trade unionists. Robinson said it was a pity the war effort had to be injured because his members had not secured recognition of their union. King told them there was no other answer he could give to them, adding ambiguously, 'I could not play two parts.' The trade unionists finally admitted 'there is no hope that the government will intervene.'[58]

The tone of the meeting conveyed by King's diary was civil, if cold and contemptuous. However, King's unusually detailed diary notation of this short and rather uneventful meeting suggests that he may well have rationalized his behaviour in his own account. Conroy later told the miners a different version of the meeting. The union delegation was granted only seven minutes with the prime minister, during which time he had not hesitated to 'allow his anti-labour feelings to show.'[59] Later in the month, on 26 February 1942, Conroy repeated this version to the CCL executive council meeting. He recounted how the prime minister was asked to bring about an honourable settlement to the dispute, but King would give no guarantee. Conroy remarked that he had never met a man 'whose attitude toward organized labour was so cold and hostile.' The prime minister apparently was so abrupt and blunt with the delegation that his secretary, Turnbull, afterwards apologized for King's treatment of the labour leaders. A CCF newspaper reported that at this meeting the labour men, after being told there would be no government intervention, in desperation asked, 'Is there no one in Canada to whom we can appeal for justice?' – at which point King is reputed to have gotten up and walked out of the room.[60]

One commentator on the labour scene noted at the time that 'one should not take too seriously the political differences between [Ontario Premier] Mr Hepburn and Mr. King': 'Hepburn ... is loudly vociferous; he shouts about the CIO; he sends his special police up in force against the wishes of the local authorities and then sends up his Attorney-General to review his troops on the battlefield. Hepburn likes to fight in the limelight. Mackenzie King moves more subtly and darkly. He does nothing. But in the end, it is Mackenzie King's deathly silence that breaks the strike.'[61] Both the federal and provincial governments contributed to defeating the Kirkland Lake strike. When it was over, the managements of the mines expressed their appreciation to Prime Minister King and to Premier Mitchell Hepburn for their roles in the strike. Management recognized that the policies of both governments had enabled them to win. Dr W.P. St Charles, president of the Lake Shore Mine, personally thanked King for his 'keen understanding and sympathetic attitude which you and the other members of your government have shown towards the mining industry in the recent trouble.'[62] There is no evidence to indicate whether King was pleased to receive such praise.

It is evident that the decision to maintain the *status quo* had not been taken lightly, but rather followed upon a careful (if incorrect) analysis by the minister of labour and a personal assessment by the prime minister himself. But whether the ultimate decision resulted from 'callous indifference, cowardly withdrawal, deliberate hostility to labour, or just chronic disinclination to act, it [was] in any case characteristic of Mackenzie King.'[63] His inclination, philosophy, and experience made him reluctant to endorse collective bargaining, and the political and economic pressures were, as yet, insufficient to force him to do so. As a result, the situation in Kirkland Lake was left to take its course.

9
The strike is lost

By 12 February 1942, the Kirkland Lake strike was lost and the humiliated miners returned to work. Despite generous contributions, the lack of finances crippled the strike. As the dispute dragged into a bitterly cold January the union leadership realized that the government was not going to intervene. The importance of the gold-mining industry to the war effort was declining, and with it the union's bargaining position. After Pearl Harbor and the American entry into the war in December 1941, it became obvious to the union leadership, particularly those with an understanding of economics, that 'our position was all but gone. Our local was sunk; it was sinking quickly.'[1] The Canadian and American governments were rapidly moving to a position where payment for war supplies would be covered by lend-lease or similar arrangements, and not by gold transfers, and the production of base metals would be given priority over gold mining.[2] The changes in American financial policy encouraged significant cutbacks in Canadian gold production.[3] Furthermore, since the American unions had adopted a no-strike pledge, the Kirkland Lake local could no longer expect significant American aid.

These international developments, the local situation, and the continued refusal of the federal government to intervene all implied that the strike could not be maintained. The union called a meeting of its local executive, which Mine Mill's President Reid Robinson and the union's international research director also attended. The situation in the union as a whole was grave and this was an additional consideration in concluding the strike: 'We knew it was gone. Where did we have any holdings? [In] BC there was nothing, and we did have something there and we lost it in 1919 and it never grew again ... Timmins had gone out of existence in the process of supporting us in the Kirkland Lake strike.'[4] It was decided to shift the union's organizational emphasis to the nickel industry in Sudbury.

A membership meeting of Local 240 was called and the miners concurred in their leaders' decision to call off the strike. Pat Conroy (CCL secretary-treasurer) was on hand to address the miners whom he said had been fighting both their immediate employers and 'entrenched capital ... of which the Canadian government was the representative.' He continued prophetically, 'This is only the first shot in a campaign that will bring freedom to the workers of this country.'[5]

On 5 February, the union wired the prime minister that the miners would return to work and begin the 'immediate resumption of full operation' if the workers were rehired on the basis of seniority and the government promised protection against company reprisals. The union contended that the mine owners were prepared to go to the 'extreme limits' to break the strike, that this intensification of the dispute would create national disunity, and that therefore the union was offering to end the strike in the interests of the war effort. The statement did not disguise the fact that, out of weakness, the union was conceding defeat. The offer was made public after the closed session with the members of Local 240. The mine operators made no comment.

The government remained consistent in its policy of 'non-interference,' in part because of political considerations. When the government received the telegram from the union, Prime Minister King directed that no reply be sent until after 4:00 PM, 6 February (which was a Friday), to ensure that the union could not receive it and act on it until after 9 February when the crucial South York and Welland by-elections would be safely over. Because labour minister Humphrey Mitchell was running for election to the House of Commons in Welland, King further suggested that Mitchell be informed of the union's communication as it 'might have a bearing on his election.' The Kirkland Lake strike and labour policy generally had been an issue in the CCF campaign against him.[6] When the deputy minister of labour, Bryce Stewart, told Mitchell about the telegram from Kirkland Lake, Mitchell responded that no action should be taken until Monday, 9 February, 'as Mr Hepburn is just waiting to denounce the government for selling out to the CIO if he can find an excuse.'[7]

The government acknowledged that it had received the union's communication, but indicated that 'no reply could be sent to the miners until the matter had been given considerable study in co-operation with labour minister Mitchell, now engaged in the by-election campaign in Welland, Ontario.'[8] As a result, labour's reaction to the loss of the Kirkland Lake strike did not influence the immediate political situation.

In the end, the government made no commitments to the union. The official view was that the miners had called a strike while the labour minister (at that time McLarty) was still endeavouring to bring the two parties together to settle the dispute. The government was convinced that the strike was over and that there was nothing to be gained or lost by further official action.[9] In any event, the union's desire for jobs and seniority might not be possible to fulfil, since the mine operators had indicated there would be fewer jobs. The government tacitly accepted that it would be the employers who would decide whom they would rehire. The government also refused to intervene to prevent the black-listing of the strike leaders, although it was requested to do so by J.L. Cohen.[10] There was now no possibility that the dispute would go to the National War Labour Board (NWLB). It was over.

Throughout the strike, and particularly toward the end, letters and telegrams from local unions all across the country urged the government to do something for the Kirkland Lake miners. The government replied only to the Nova Scotia coal miners, and the Sydney and Sault Ste Marie steel locals, which had requested SWOC Director Charlie Millard to declare a national shutdown of all the basic steel plants in support of the Kirkland Lake miners. Mackenzie King told those locals he must confer with the minister of labour and his Cabinet about the situation.[11]

Most of the other correspondents received a form-letter reply. On 2 February 1942, at the end of the dispute, the prime minister's principal secretary was instructed to reply that

P.C. 2685 is a declaration of principles for the avoidance of labour unrest during the war suggested to both employers and employees by the government. Certain principles would require changes in attitudes and procedures by employers while others would similarly affect employees. The government has not compelled acceptance of these principles by either of the parties. Although the first principle states that every effort should be made to speed production by war industries, the government did not exercise compulsion on miners involved in the slowdown in Nova Scotia coal mines. Order-in-council also states there should be no strikes or lockouts but strikes have occurred without application of penalties. Government's view is that if policy is enforced on Kirkland Lake operators it must also be enforced against employees who do not conform with policy. *Government has no present intention of resorting* to compulsion in these matters.[12]

This statement accurately portrayed the government's policy throughout the conflict.

On 11 February, a union statement was issued officially declaring the end of the strike and expressing the hope that the government would co-operate in assisting the miners to return to work. The mine operators had succeeded in maintaining an open shop in the mines, but the union hoped that they would rehire in accordance with seniority, as quickly as possible, so normal community life could be restored. The hope was in vain. Management reopened the mines in their own way.

The union's defeat was complete as management reports filed with the Department of Labour indicated.[13] There were no terms of settlement, no written agreements, no union recognition. The wage rates were the 'same as before'; the hours of work were the 'same as before' – forty eight hours. The Macassa mine management even added that terms of a settlement were 'under consideration with [an] elected employees' committee.'

Production was resumed on a normal basis in all the mines between 12 February and 21 February 1942, although the size of the work force was reduced by 25 per cent. In each mine a committee of workers was elected to deal with management regarding grievances and questions of working conditions.[14] It was the same pattern that had occurred in so many previous mining disputes that the union had lost. On 16 February, the union distributed a leaflet entitled 'Remember Kirkland Lake.' The leaflet expressed the local's deep bitterness:

The historic strike at Kirkland Lake is over. The men who have carried the banner of trade unionism and freedom through all these long months are going back to work right now without recognition or anything else. Once again, the vicious coalition of mine management, Canadian vested capital, the government which is not the people's and a police army of occupation have starved out a union of Canadian workers and maintained the feudal open shop. Once again, the boot heel has descended on democracy. Abroad we call this fascism ... Yet these courageous men are not going to be defeated. A skirmish lost – a battle yet to be won. And won it will be, for these men held out long enough to educate a great deal of Canada.[15]

The immediate problem for the union concerned the companies' punitive rehiring policies. When the strike ended, many of the older miners and the 'new Canadians' were not rehired. But the young Canadians did receive notices: 'A job is waiting for you.'

Bob Carlin (financial secretary) informed the local executive of this situation and the local held a meeting to discuss it. At the meeting, one of the young Canadians, Larry Sefton, stood up and said, 'I'll be Goddamned if I'll go back before the old fellows.' All the other men agreed and the meeting adopted that policy. For several months many of the young men stayed around after the strike 'and the young guys would smile at each other – "there's still some old fellows not rehired."'[16]

No trade union officials were rehired. The local president, William Simpson, was fired by the Lake Shore mine. The company claimed he had not been employed since 16 November 1941 – two days before the strike – and had been fired because of 'lost time.' Joe Rankin (local vice president), in a wire to McLarty, charged that this was an example of 'direct discrimination.' The cause of the 'lost time' was the conference in Ottawa urgently requested by the labour minister himself immediately prior to the strike. Rankin urged that the Department of Labour should act immediately to compel the company to rehire Simpson, 'and that the Department should oversee the return of the men to work.'[17] This indignity towards the local union president was in contravention of PC 2685 and section 502A of the Criminal Code. Humphrey Mitchell said he would give consideration to this matter but had no immediate comment. Ultimately he advised that union complaints were being referred to the 'mining authorities,' but even this was not done, nor did the Industrial Disputes Inquiry Commission (IDIC) investigate this example of discrimination.

The union informed the government of numerous instances of discriminatory hiring practices. The issue was raised in Parliament by CCF members and on one occasion by Mr Hanson, the Leader of the Opposition. On 13 February CCF MP Angus MacInnis asked acting labour minister McLarty (Mitchell, although a cabinet minister, was not yet seated in the House) whether the government had raised the issue with the mine operators. McLarty replied vaguely that he *thought* the government had requested them to be magnanimous. There had been no direct intervention and the mine owners had given the government no formal indication that there would be no discrimination. In any event, McLarty continued, there had been no complaint (an unusual statement in view of the unions' telegrams). CCF leader M.J. Coldwell, sensing the ambiguity of his answer, questioned him again. Had the matter been taken up? McLarty could not answer definitely and asked that the question stand so he could confirm what had been said. He minimized the situation and again repeated that he was not aware of any complaints. 'I understand the men are going back to work but no discrimination has I believe, so far been shown.'[18]

Ten days later MP Clairie Gillis once again raised the situation in the House and asked what the government was doing about it. Mitchell (by then an MP) replied that he had made inquiries and had, that day, received the following reply from the Kirkland Lake committee of operators: 'We are taking men back as quickly as possible with due regard to balancing our operations, but it is slower than we anticipated and we are giving preference to the older and married men first and I am sure that when the matter is finally cleared you will be satisfied that we have done a good job on it.'[19] Both Angus MacInnis and Unity MP Dorise Nielson, who had returned from Kirkland Lake the day before, considered this statement 'an absolute falsification of what is going on.'[20] When Gillis pressed Mitchell about incidents of rank discrimination, Mitchell replied that there were problems because there were more miners than could be absorbed by the mines (which was true, as the mines had decreased production and the size of their work force) and that, in any event, the men 'returned unconditionally without intervention from the Department of Labour.'[21] After meeting the union delegation following the strike, the minister had complied with their request and wired the operators' committee requesting a non-discriminatory policy. The mine operators replied that they were following such policy 'as closely as possible and having due and proper regard for the efficiency and safety of the operations.'[22] Mitchell accepted this statement as true, notwithstanding the reports of various MP's and union complaints to the contrary.

By the end of February, about 500 men had been taken back to work, but no union officials were among them. At Teck-Hughes, only seven of the previous work force of 400 were recalled. Part of the mine was closed and men with seventeen to twenty years of service were refused jobs. Approximately 1800 men remained idle, and some 700 had left town.[23]

The older miners suffered the most. They were the last to be hired and the first to be laid off if the production demands so required. These men stayed in Kirkland lake waiting for work and became a group of penniless unemployed. The union had given unemployed miners relief in the immediate post-strike period, but by 13 March 1942 it could no longer continue to pay strike benefits and requested Teck Township to undertake the payment of relief to these families. The council was in a difficult financial situation and approached the province, but without success. The Honourable Farquhar Oliver, minister of welfare in Ontario, said the Kirkland Lake miners who had been on strike and had not been re-engaged were not eligible for relief so long as they were employable. MPs Gillis and Neilson raised the matter in the House and again urged federal government action, but without success. Altogether 1000 men were not rehired and all of them were 'perfect specimens of physical fitness.'[24]

Once the government had classified gold mining as a 'non-essential' war industry (in 1942) production was further reduced and thereafter there was no hope of these men being re-employed in the gold mines. The steel industry was given top priority in the war economy. Though offers were received to hire Kirkland Lake miners in the steel mills of Sault Ste Marie, only the younger men were transferable. The Leader of the Opposition, the Honourable R.B. Hanson, pointed out that if they could not be employed in Kirkland Lake, the position of most of the unemployed miners was hopeless.[25] Again the government's inactivity was criticized – this time by the Conservatives. Mitchell suggested that transfers of miners to other industries could be arranged by the Employment Service of Canada, the agency responsible for manpower relocation; but he added that the employers were anxious that the men be available to return to the mines if and when they were needed. The government expressed little interest in actively searching for new employment for the miners, even though there were manpower shortages and their skills could have been profitably employed. The union charged that the mining companies wanted to maintain an over-supply of labour in the vicinity of the mines in order to combat unionization of the work force, and complained that government inactivity was bolstering the employers' anti-union position.[26] At the time the mine operators were actively engaged in establishing employer-dominated employees' committees.

Mitchell opposed the union to the end, even in the matter of transfers. In November 1942 the resignation of Elliot M. Little as director of the Selective Service Commission was partly due to Mitchell's rejection of his plan to relocate gold miners in base-metal mines that were more essential to the war effort. Mine employers were opposed to this policy because they did not want to hire men sympathetic to unions.[27] Mitchell was not prepared to press the issue, and the Kirkland Lake miners remained idle. The union had lost and the government saw no need for any magnanimous gesture.

In a broader sense, the Kirkland Lake strike was a turning point. It marked the peak of labour's frustration with an insensitive and unsympathetic government which was, by this time, regarded as actively anti-labour. This mood was expressed by Angus MacInnes (himself a trade unionist from British Columbia) in a speech in the House of Commons, delivered on 5 February 1942: 'As anti-social employers gloat over victories gained over their workers, with the assistance of a servile and anti-labour government smoldering bitterness grows in the hearts and souls of beaten and sullen workers.'[28] The union, the CCL, the individual locals of both the CCL and TLC, the church groups, and the 'Kirkland Lake Committees' across the country had all been beaten. But the Kirkland Lake strike had raised fundamental issues for the labour movement. It had aroused intense feelings. And for the first time during the war, the movement was united in a common national concern.

10
The effects of the
Kirkland Lake strike

THE GOLD-MINING INDUSTRY

The Kirkland Lake strike affected the gold-mining industry, the town, the local union, the labour movement, and political events. The Ontario Department of Mines reported that in 1941 the Ontario gold-mining industry had experienced one of the most turbulent years in its history. Total output was down 18 per cent from 1940.

For the first time in thirteen years, output had not increased, but the strike was probably less significant than the changing world market conditions after the United States entered the war and greater priority was given to base-metals production.[1] An increase in production in the Porcupine, Larder Lake, and Northwestern Ontario camps partially offset the Kirkland Lake losses.[2] In addition, the mining companies had taken specific action designed to reduce the effectiveness of the strike. During the period of enforced conciliation they had built up a stockpile of ore that continued to be milled after the eventual, inevitable, breakdown. The companies also continued to operate the mines with 'scabs' and managerial personnel. This continued operation had a significant psychological impact, but its contribution to output was probably less than that of the stockpile. During the strike the owners concentrated their efforts on mining and refining the higher grade ore, which had a higher market value. As a result, the production and profit position of the companies was better than might otherwise have been expected.[3]

Prior to the strike, the *Northern Miner* had reported that 'unquestionably the labour dispute instituted by the CCL [Canadian Congress of Labour] has hurt mining stocks. It couldn't help but do so.'[4] However, stockholders with investments in the gold-mining industry maintained their confidence throughout the strike. For the most part they did not wish to liquidate their holdings, since this would signal a loss of confidence in the industry and might cause a sharp decline in stock prices. While they were apprehensive about the short-term results of the strike, they had no illusions about its final outcome. Most of them did not sell their stock during the strike,[5] and the impact on the value of mining stock was marginal.

The mining industry had claimed that a strike, or increased wage costs, would permanently damage its economic position and might ruin the marginal producers. In the period following the strike, production was reduced – the value of daily production declining by $200 000 as a result of a reduction in tonnage milled.[6] Two mines were closed, and three mines did not return to their pre-strike levels of production. When the mines had re-established their businesses at this new level of production, there were 1000 fewer employees.[7] Nevertheless, the decision to cut back production created a situation in which the *per unit* profit actually increased, so that the economic impact on the profitability of the industry was not as serious as many had expected.[8]

The 'boom period' of the gold-mining industry had passed, and the strike was used as a reason for retrenchment in the industry, but the stockholders were not seriously affected. The main effect of the cut-backs was to disrupt the economic base of the town and to create great hardship and upheaval for those miners who were not rehired. In terms of human suffering the aftermath of the strike was serious.

The decline in production in the industry was blamed entirely on the strike. One newspaper commented that it was 'unfortunate that the net result of the strike to date may be a reduction in total employment.'[9] Of course, significant changes in the international gold market had also influenced the business decisions that had been taken. Much of what occurred would have happened in any event. The strike merely added to the pressures of the market, inducing management to cut the size of the work force, to abandon further exploration,[10] and to discontinue the mining of low-grade ores, which were becoming increasingly less profitable. Nevertheless, the timing of the strike and the subsequent cut-backs created the myth that these changes were solely the result of the strike. This myth was perpetuated by the industry and its press and was apparently widely believed by the townspeople.

THE COMMUNITY

The impact of the strike on the community of Kirkland Lake is difficult to evaluate. During the strike, the *Timmins Press* reported that Kirkland Lake, once 'one of Ontario's most hustling communities,' had suffered a slump because credit was restricted.[11] Wholesalers advised merchants that goods sold previously on a two-week or monthly credit basis were henceforth available only for cash. Some of the merchants adopted the same rule in their retail trade. Others continued to give credit until the miners received their last pay cheque on 25 November. Some merchants contended that Kirkland Lake had been in a slump ever since 1500 young men had enlisted in the armed forces, and the strike merely aggravated the situation. Because of the long period of apprehension prior to the actual strike, the miners had saved their money and even during this period consumer spending in the town was probably reduced. There is no doubt that the economic effects of the strike were felt in the small business community, which depended upon the miners' spending.

The mining companies' decision to reduce employment and production totally disrupted the economic life of the town. Between eight and nine thousand people moved away. As one striker commented, it was fast becoming a 'ghost town.'[12] Many of the young unemployed left Kirkland Lake permanently. The older men stayed and remained unemployed. For people who remembered the vitality of the town during the 1930s, the situation was particularly depressing.

Following the strike, municipal tax revenue derived from the mines declined by $93 000. The Teck Township Council responded to this situation by seeking a moratorium on home loans and mortgages. Expenditures were reduced by $83 000 below the level of the preceding year. In March 1942, there were six to seven hundred vacant homes and apartments. As its size decreased,[13] the town also lost per capita provincial street and highway subsidies.

The loss of the strike and the subsequent unemployment and reprisals did little to improve social and community relations. There continued to be animosity between union and non-union men, particularly since some of the employee committees were re-established after the strike in order to compete with Local 240 and frustrate any efforts to reorganize the local. There was lingering bitterness between the union and management. The mine operators remained completely opposed to any form of independent unionism anywhere in Northern Ontario. At a meeting of the Sudbury Rotary Club, a representative of the Sudbury Business Men's Association recounted sensational stories about the Kirkland Lake strike and the local mine managers reaffirmed their determination never to recognize the CIO.[14] When Mine Mill (IUMMSW) switched some of its activity to the nickel centre, Sudbury citizens were urged to fight the union and avoid a repetition of the problems that had arisen in Kirkland Lake.

These were not the only views taken of the strike, however. Many Kirkland Lake citizens felt that the strike might have been averted had the companies been more conciliatory.[15] They remained sympathetic to the union's position and later supported both CCF political activities and the union's attempts to reorganize the mines.

It is evident that it was not just trade unionists who were urged to 'remember Kirkland Lake.' As late as 1946, during the Hamilton steel strike, the Kirkland Lake operators' committee wrote to H.C. Hilton, the president of the Stelco works, saying that 'the strike situation at your plant parallels our own experience of 1941–42 so closely that, in the interests of the *common cause*, we felt impelled to write to you.'[16] Ultimately labour relations in the northern gold-mining industry were stabilized and because of the existence of positive legislation, which compelled employers to recognize unions with majority support, by the 1950s collective bargaining was established. In 1946, however, management was still hostile to organized labour, and the town felt the repercussions of that labour/management antipathy.

LOCAL 240

For Local 240 the strike was disastrous. The local incurred a $26 000 debt.[17] Some $12 000 was owing to local merchants who, after a meeting with the union, agreed to continue to give credit to men taken back to work. To help to alleviate this financial situation, the National Kirkland Lake Committee appealed to other unions to continue sending money despite the failure of the strike, and for some time funds continued to be sent.

Efforts to solve the financial problems also involved criticism of the CCL's role during the strike. SWOC Director Charlie Millard believed that the congress had a continuing moral obligation in Kirkland Lake since the outstanding debts of Local 240 could be demoralizing to both the miners and other struggling locals. Millard argued that the failure of the strike was the fault neither of the union,[18] nor of the unions that had given assistance. He recommended that the CCL give an additional $2000 and that $5000 be requested from the international office of the Mine Mill union.

Pat Conroy, CCL secretary-treasurer, was well aware that there was criticism of the CCL's role in Kirkland Lake, and that some trade unionists felt that the CCL could have done more. However, Conroy was also aware that many of the unions that were most critical of the congress had themselves contributed little to the strike.[19] Accordingly, the congress rejected Millard's proposal and undertook only to remind its affiliates that the fight was not finished and that continued support was necessary.

The financial problems of Local 240 were not finally resolved for almost two years. In December 1943, the international union provided $29 000 for the discharge of the outstanding debts of its Kirkland Lake local. The final cheque was delivered to William Simpson, the union's international board member in Toronto, by secretary-treasurer Jim Leary of Denver. Simpson paid the Teck Township Council and the Kirkland Lake merchants for supplies advanced during the strike. It was hoped that the discharged debt would 'end propaganda the debt made available to those opposed to the union and to recognize also the outstanding part the Kirkland Lake miners had played in bringing organized labour to the realization of necessary action to remedy its [legislative] position.'[20] The payment did enhance the position of the local in the town,[21] and of the union in Northern Ontario generally. Since Mine Mill was still actively trying to organize in that area, it was important to maintain its reputation.

Many of the local's members remained unemployed after the strike. The local did what it could to assist them to find alternative job opportunities and solicited information in this regard from the CCL and from local labour councils throughout Ontario.[22] Ultimately, however, the successful relocation of large numbers of unemployed miners depended upon the co-operation of the federal government which, during the war emergency, exercised responsibility for manpower adjustments. Despite manpower shortages, the federal government did little to assist the unemployed miners.

Local 240 itself was in disarray. The union hoped that members who stayed in Kirkland Lake would remain loyal; and a meeting held several weeks after the strike was attended by 400 members. Nevertheless, the local temporarily ceased to function as a bargaining agent for its members, and the fortunes of Mine Mill seemed to be on the decline again. The union shifted its concern to Sudbury, where some preliminary organizing activity had already begun. Organizer Bob Carlin went to Sudbury, but the campaign was already facing violent opposition. Forrest Emerson, an international representative of the union, and Jack Whelan, an international organizer, were beaten up and the union office was completely demolished by intruders who were apparently looking for the union membership lists.[23] The attack was seen by the union as the inevitable result of the government policy in Kirkland Lake, as apparently anything could be perpetrated against the union with impunity.

A committee of Sudbury miners informed the federal government of the attack and of an 'incessant anti-union campaign' by the *Sudbury Star*. The matter was even raised in the House of Commons. Labour Minister Humphrey Mitchell advised the House that he had no information about the matter, and that, in any event, the responsibility for maintaining law and order was provincial and municipal. Justice minister St Laurent added that the RCMP (Royal Canadian Mounted Police) had no knowledge of the incident, and therefore there would be no investigation.[24] Given the expressed opposition of the provincial government to unions in the mining industry and the role of the police in Kirkland Lake and elsewhere, the employees expected and received little sympathy from that quarter.

The local was weakened by the continuing surplus of labour in the area, and the companies' efforts[25] to establish company unions. Employers usually are staunch advocates of the 'right to work' and resolutely oppose any form of union security – particularly the 'closed shop' or 'union shop.' Nevertheless, one company established a compulsory dues check-off system for *its* employee association, and made it a condition of employment that a worker become a member of the association after six months probation. The contract that included these clauses had not been ratified by the miners. It was imposed. The majority of that particular work force had remained members of Local 240 and eventually took a representation case before the Ontario Labour Court.[26]

After the Kirkland Lake strike Local 240 lost its most active supporters. They were never rehired, and were forced to seek work in Sudbury and southern industrial centres. Nevertheless, over the next few years, after PC 1003 was proclaimed in 1944, the local slowly gained recognition on a mine-by-mine basis. By that time the gold-mining industry was less prominent, the work force was smaller, and industrial unionism was firmly established. By the end of the war the international union was more interested in base-metal mining, internal union political differences,[27] and the threat to its jurisdiction from the United Steelworkers of America.[28] But Local 240 continued to organize and by 1945 had negotiated contracts with Macassa and Kirkland Lake Gold mines. In 1944 it lost its appeal to represent miners at the Wright-Hargreaves and Sylvanite mines. It had won the support of a majority of those voting at each mine, but since it did not win a majority of those *entitled* to vote, it was not certified. In 1945, such cases provided the basis for a CCL appeal for amendments to the collective-bargaining order, PC 1003. Its certification at Lake Shore mine was finally upheld in 1946 (after an appeal by the Independent Canadian Mine Workers Union).

THE LABOUR MOVEMENT

The entire labour movement was worried by the outcome of the Kirkland Lake strike. The trade unionists who headed the CCL unions, and who had been closely involved, were desperate over the loss and over the state of legislation affecting unions. Institutionalized delay and the calculated inactivity of the government blunted the effectiveness of the strike weapon, and implicitly supported employers. The united strength of the labour movement had been mobilized to win the Kirkland Lake strike and it had failed. There was concern about the future.

One indication of labour's uneasiness was a remarkable exchange of letters immediately following the strike. Union President Reid Robinson, after conferring with Pat Conroy, other Canadian trade unionists, and the union's lawyer, J.L. Cohen, wrote to Philip Murray, president of the CIO, to ask that he and President Green of the rival AFL (American Federation of Labour) come together to co-operate (as they had in a war Victory Program Committee in the United States) to 'work out a unity legislative program that we could apply in Canada.'[29] Such assistance was to be informal and unofficial. J.L. Cohen, who understood the politics of the CCL, and the need to approach the American union officials carefully so as not to jeopardize CCL independence, wrote to Robinson: 'I agree with the general course outlined, but cannot urge too clearly the necessity of avoiding any "official" action along the line suggested until the whole situation has been "unofficially" canvassed and agreed upon by the interested parties.'[30] Pat Conroy, who also agreed with the proposal, was to be the liaison person in Canada and was to establish co-operation between the CCL and the TLC (Trades and Labour Congress).

It is clear that the Canadian trade unionists were seeking help, and that Reid Robinson was trying to salvage something from the strike to enhance the prestige of IUMMSW. The result was a proposal that in ordinary circumstances would have been extraordinary – especially given the rivalry between the AFL and the CIO. Robinson's deep concern with the government's role was apparent:

In all my experience I have never encountered so much indifference on the part of governmental officials towards the rights of workers ... We feel that if you and President Green were brought together with representatives from every section of the Canadian labour movement at a legislative conference to be held in Ottawa, that very quickly the desired results could be obtained. I am sure that through your present contacts with President Green that such a conference could be arranged and the presence of yourself and President Green at such a conference would have a very important effect upon the entire Canadian nation.[31]

Conroy also wrote to Philip Murray to seek his support.

This proposed conference in Ottawa did not take place but the idea itself, and the CCL's involvement with it, was an indication of the level of frustration which Canadian trade unionists felt after the loss of the Kirkland Lake strike. No more pressure could be exerted in Canada than had already been mobilized: trade union solidarity in the industrial sphere had proved inadequate. Bringing in prestigious American union leaders who had gained the ear and respect of their own government seemed a useful approach to increase pressure on the Canadian government for legislative change.

Despite the recriminations following the strike, its ultimate impact on the Canadian labour movement was mixed. Organizationally, the labour movement may even have benefitted. The mine operators' refusal to rehire active trade unionists meant that such persons were forced to move on and seek work elsewhere. Many of the younger miners were forced to go to Southern Ontario and find new jobs in Toronto, Hamilton, and Oshawa. Some were disillusioned and bitter about their experiences in Kirkland Lake and did not immediately get involved in union activity. When Larry Sefton was approached by Charlie Millard and asked to join the staff of the Steelworkers' Organizing Committee (SWOC), he refused at first because he 'had had a bellyful'[32] in Kirkland Lake. Ultimately, however, he and others joined the expanding staffs of the new industrial CCL–CIO unions. They had not abandoned their idealism and their experience from the Kirkland Lake strike was an asset. They had, after all, learned how to organize and deal with the complexities of labour relations under siege. As SWOC representative Eamon Park said in later years: 'In the days right after the Kirkland Lake strike, if you were going out organizing, you first found out where the Kirkland Lake boys were because you immediately had a good local union committee because they'd come through the mill in every sense of the word.'[33]

These ex-miners used their talent to organize local unions in various industries. Some went on to assume positions as staff representatives of the new unions or even higher leadership positions. Jim Russell, Joe Rankin, Jock Brodie, and Bill Sefton became international representatives on the staff of the United Steelworkers of America (USWA). Eamon Park became an international representative and subsequently assistant to the Canadian national director of the USWA. Larry Sefton, the young recording secretary of Local 240, went on staff as an international representative of the steelworkers, and in 1953 was elected as director of District no. 6 of that union. He later became a member of the international executive board of the Steelworkers' Union and a vice-president of the Canadian Labour Congress. Bob Carlin, Local 240's financial secretary, became the Canadian representative for District no. 8 and in 1942 a member of the international board of the IUMMSW. In 1943 he was elected to the Ontario legislature as the CCF member from Sudbury. William Simpson, president Local 240, became a staff representative for the IUMMSW. 'Doc' Ames, who on the eve of the Kirkland Lake strike was a first aid man working at Lake Shore mine and sided with the miners, emerged from the strike to begin his life's work as a political organizer for the CCF–NDP. The Kirkland Lake strike politicized these miners and propelled them into careers as trade unionists and political activists.

The labour movement itself recognized that it had been badly defeated at Kirkland Lake, but it could still see some benefit from the efforts to organize and maintain the strike. *Steel Labour* commented: 'The Kirkland Lake strike is over. There is no point in denying that it has been a defeat for organized labour in this Dominion ... There are, however, rays of light shining through the sombre picture; one has seen the splendid unity displayed by the Canadian labour movement in connection with the Kirkland Lake issue.'[34]

Parts of the national organization remained intact and continued to function long after the end of the strike. In some centres like Sault Ste Marie, a CCL labour council was established for the first time. As the *Canadian Unionist* wrote, 'The whole idea is an outgrowth of the Kirkland Lake committee meetings.'[35] Organized labour had gained and maintained the support of new sections of the public, particularly among the churches: 'A fair job was done in influencing public opinion to the degree that some churches, and persons outside the movement and even some city councils rallied to labour's side in condemning both operators and government.'[36] The churches generally supported the union's position on the issue of recognition, and despite the length, cost, and ultimate loss of the Kirkland Lake strike, this support was undoubtedly a factor in achieving the important legislative advances that were made in the next few years.'[37]

On 26 February 1942, two weeks after the end of the strike, the executive council of the CCL prepared a memorandum to the government and presented it the next day. It outlined in detail all of the matters since the war began, which made organized labour 'frankly apprehensive of the attitude of the Government toward Labour organization' and which 'caused discontent and irritation among the workers.'[38] At the top of the fourteen-point list of grievances was the failure to enforce order-in-council PC 2685 (see appendix). The brief made three specific recommendations: 1) that there be adequate representation of organized labour on government bodies; 2) that a comprehensive labour policy be enacted guaranteeing employees the right to organize and bargain collectively through the union of their choice; and 3) that a wage policy be established under which wages and working conditions would be determined by collective bargaining, subject only to such controls as *might* be considered necessary by industrial councils.[39]

One theme pervading the entire brief was the contrast between allied countries like Great Britain and the United States, where labour had been taken into full partnership in the war effort, and Canada, where the 'direction and control of the war production effort [was] left entirely to industrial and financial executives while government employees and the workers of the nation are treated as mere cogs in the machinery of production rather than as human beings with as much stake in the war as any other class in the community.' This alienation at the national level was clearly intensified by the experience of the miners in Kirkland Lake. In labour's view, the failure of the government to give sufficient recognition to workers' organizations had serious effects on morale, productivity, and industrial relations. It was obvious that the government had no intention of enforcing its labour-relations principles. Protests had been futile. The government's expressed aversion to 'compulsion' was mere sophistry. The government had exhibited no such qualms when it imposed compulsory conciliation, compulsory strike votes, and compulsory wage controls. As the congress concluded:

This is specious reasoning, since all legislation is compulsory, and reference has been made herein to a number of items of legislation which compel the workers to act or refrain from acting in certain ways. The fact is, of course, that the Government's so called Labour policy is not legislation at all, although it is embodied in an order-in-council; it is nothing more than an empty gesture, a 'recommendation' which the Government will not even follow itself, in dealing with employees in Government-owned and controlled plants.[40]

The CCL memorandum included a detailed accounting of the wrongs allegedly committed in Kirkland Lake. This conflict was regarded as but the latest example of the government's refusal to enforce the principles that it purported to approve. The miners had gone on strike to obtain rights that the CCL regarded as basic in a democratic system. The failure of the government to protect those rights had engendered resentment and hostility among large numbers of workers all over Canada[41]: 'The Congress feels that the Government should be aware of the seriousness of the situation which has developed. Furthermore the workers are determined to keep up the struggle for union recognition and collective bargaining as long as that may be necessary.'[42]

After meeting this obviously angry labour delegation, Mackenzie King commented in his diary: 'As I looked into the faces of the men there, they seemed to be those of men who for the most part were feeling that the whole of society was against them. That they were full of grievances and full of suspicion, and not believing anything that was being told to them.'[43] Unwittingly perhaps, the prime minister himself, in the privacy of his own counsel, had described the legacy of his government's labour policy.

King was disturbed by the tone of the delegates, but attempted to pacify them. After the meeting the prime minister went outside to be photographed with the delegation – in a gesture intended to ingratiate himself and which they apparently did appreciate. He made a revealing comment in his diary about the meeting: 'I felt sorry not to have a little more time to speak to them carefully. They represent a branch of labour which is having a hard time. Unfortunately, under the influence of the worst elements of the CIO, a difficult lot to deal with at this time.'[44] King's estimation of these young, aggressive, industrial trade unionists accounts, in part, for his government's inertia in presenting legislation that would be favourable to their new unions. King did not like them. Nevertheless, he was astute enough to realize that he was politically isolated from them, and that this might, at some point, be inadvisable. In the short term he did nothing; but having recognized the problem, in his own good time, he began to seek acceptable solutions to some of their grievances. When labour openly supported the CCF in 1942 and 1943, an increasingly concerned prime minister was finally moved to respond.

The 1942 TLC brief to the government also called for the enforcement of PC 2685. The TLC was more vigorous than it had been in previous years. Although it was not as openly critical as the CCL, the TLC brief reflected the growing disaffection of all segments of the labour movement. The brief demanded that the Cabinet immediately declare that the PC 2685 principles would govern labour/management relations in Crown corporations. The Cabinet indicated agreement with the TLC proposal and promised immediate action, but despite this promise no order was forthcoming until December 1942. The order was delayed in part because of the absence of ministers C.D. Howe and Humphrey Mitchell, who were in Europe.[45]

The government did make some minor changes following its meetings with the two labour congresses. In March 1942, it abolished the NLSC (National Labour Supply Council) and restructured the Interdepartmental Committee on Labour Co-ordination so as to include one labour and one management representative. This action was the response to criticism concerning the absence of labour representation on government boards and committees;[46] however, the government was not prepared to significantly alter the manner in which decisions were made. The reconstituted committee merely became less influential. In the meantime there was more criticism emanating from the 1942 conventions of both labour congresses.

THE LABOUR CONVENTIONS, 1942

An atmosphere of bitterness and mistrust was evident at the CCL convention held in September 1942. The verbatim transcript of the proceedings reveals that six months after the defeat in Kirkland Lake, the issue continued to dominate the labour scene. When the federal and Ontario ministers of labour were introduced as guest speakers to the convention, delegates called out from the floor, 'What about Kirkland Lake?' and 'Remember Kirkland Lake.'[47] Numerous allusions to the strike were made from the platform and the floor.

In its report to the delegates, the CCL executive board put the 'blame' for the loss of the strike squarely upon the shoulders of the government, the prime minister, and the new minister of labour, the Honourable Humphrey Mitchell who, as chairman of the IDIC, had previously recommended a 'company union set-up.' The report thanked the delegates, non-labour groups, and individuals who had publicly supported the strike, and predicted that 'the resentment shown all over Canada against the government on this issue would have been strong enough in peace time to have brought about a change of government. The fight of the Kirkland Lake miners has not been and will not be forgotten.'[48] However, the CCL executive report did not urge immediate action and quoted an editorial in the CCL publication, the *Canadian Unionist*, to the effect that in the interest of the war effort, 'the righteous wrath of those who feel that the government has betrayed the Kirkland Lake miners must be stored up until the time comes when it may be displayed without creating more disunity.'

There was considerable animosity directed against Mitchell personally, for as chairman of the IDIC and author of the Kirkland Lake Formula he had been instrumental in promoting the policy of inaction that had contributed to the defeat in Kirkland Lake. The convention debated a resolution calling for the replacement of Mitchell in the interest of harmony and industrial peace in Canada.[49] The resolution contended that he had been a party to smashing the Kirkland Lake strike and had lost the confidence of organized labour. The CCL executive recommended non-concurrence, but the resolution created an uproar on the convention floor and was applauded by many delegates. Delegates speaking at the microphones were angry. One declared, 'We know his attitude toward the Kirkland Lake strike.' Another speaker from PWOC (Packinghouse Workers' Organizing Committee) reminded delegates that Mitchell had recommended company unions as a means of settling disputes. He concluded by saying: 'We stand by labour men who stood by us in the Kirkland Lake strike, and we repudiate labour men who shied away from us at the Kirkland Lake strike.' A steelworkers' delegate argued that if the convention failed to censure Mitchell, it would be a tacit endorsement of his policies. This delegate acknowledged that a vote of censure might weaken support within the government for legislation, but stressed that there was, in any event, little evidence of government support or even good will. He argued that legislation would be forthcoming; 'not because the powers [that] be willed it, but because the trade union organizations are growing in strength and preparing to assume their rightful place in the economy of this country.' He, therefore, opposed the executive's recommendation of non-concurrence and urged that the resolution be referred back to the CCL executive so that appropriate language of condemnation could be drafted. Another delegate said: 'When we consider his actions in the Kirkland Lake strike, actions well known to all of us, it makes a very sorry mess indeed. We want a Minister of Labour who will not double-cross us behind our back.'[50] This delegate received a round of applause.

With the convention's anti-Mitchell emotions running so high, the amendment to refer the resolution back was carried. Mackenzie King's hopes that his new labour minister would build closer links between the government and organized labour were defeated. Mitchell was also under attack by the CCF. Because of opposition from organized labour, the press, and the CCF, King seriously considered removing Mitchell from the labour portfolio.

The delegates adopted the position set out in the earlier CCL brief and repeatedly attacked the government's failure to enforce PC 2685, calling for the passage of a Canadian Wagner Act. The congress executive was instructed to prepare draft legislation along these lines and press the government to enact it. Affiliated unions were to receive copies of the draft act and take 'the most advantageous action ... to enlist the full support of the public for its implementation.'[51] Finally, the convention called on the federal government to set an example to employers and promote the war effort by immediately engaging in collective bargaining with the employees of government-owned-and-operated plants.

The resolution at this convention frequently referred in their preambles to the Kirkland Lake dispute. One interesting series of resolutions dealt with the relationship between the congress and its affiliated unions. Because of the defeat at Kirkland Lake, many delegates believed that it was necessary to have a more centralized labour movement. One resolution urged the CCL to adopt a policy of collective support of affiliated unions to the extent of calling for national or regional strikes where it was deemed necessary. Various spokesmen felt that the defeat in Kirkland Lake had occurred because the union had struck alone. Pat Conroy pointed out that these resolutions substantially increased the powers and responsibilities of the CCL and he opposed them. So did the congress executive, which, of course, was largely composed of the senior officials of the affiliated trade unions. Nevertheless, there was enough support for the resolution on the floor of the convention that it was referred back to the executive for revision.

Another resolution dealt with the method of mobilizing financial support among unions that were in sympathy with the aims of a particular strike. This resolution pointed out that most chartered and affiliated unions had contributed voluntarily to the support of the Kirkland Lake miners, but this had not been sufficient to ensure a victory in the face of business and government opposition to that strike. The supporters of the resolution called on the executive of the CCL to impose a 'tax' on member unions if financial assistance were required by *any* union on a legal strike in the CCL jurisdiction.

The debate was very emotional. A miner from Nanaimo, whose local in the UMWA (United Mine Workers of America) had supported the Kirkland Lake miners to the extent of $9000, declared: 'The Kirkland Lake fight was our fight and that of every other red-blooded union in this country. We lost that strike and it was a legal strike and a strike that was right and just.'[52] He believed that a question of national importance, such as the Kirkland Lake strike, should not be dependent on voluntary contributions. However, the executive's recommendation of non-concurrence with the resolution ultimately carried. The majority of the delegates agreed with Pat Conroy, who spoke to the resolution saying 'a strike cannot always be settled favourably no matter how much money is poured into it, and the Kirkland Lake strike was an excellent example of that.'[53] Conroy claimed that the CCL had raised $150 000 from its affiliates – 'a most magnificent indication of where labour stood on that matter, greater than ever shown in this country.' He then suggested that the debt from the strike should be paid – though not by the congress.

A.R. Mosher, speaking on behalf of his union, the CBRE, also opposed a financial assessment. He argued that his union would never allow the CCL to compel support for any strike unless the union membership itself first approved such support. This was the traditional view of the relationship between the congress and its affiliated unions. That the CCL president should be the person to espouse this view is indicative of how thoroughly accepted it was by most affiliates.

The convention rejected the idea of a CCL tax, but agreed with Delegate Dalrymple, Sr, from the steelworkers' union, that all delegates should go back to their unions and explain the financial plight of the Kirkland Lake local: 'We know the Kirkland Lake strike was worthwhile. Let us see that our membership ... gets together and wipes out this debt. We will pay our debts and continue the struggle.'[54] These resolutions calling for a change in the structure of the CCL were indicative of the general frustration following the defeat in Kirkland Lake. It was recognized that a 'just cause' was not enough. Better organization was essential.

Both the federal and Ontario labour ministers addressed the convention. Immediately prior to Humphrey Mitchell's address, delegates 'began pounding the tables and crying "Kirkland Lake."'[55] Mitchell was met with mingled booing and applause as he rose to speak. He referred to himself as a trade unionist, but gave the kind of speech that this convention did not want to hear. He said nothing new. He acknowledged that both the TLC and CCL conventions had criticized the government's wage and labour policies, but then referred to the American situation: 'The other day Mr. [Philip] Murray gave a frank message to American labour. It was that there should be no strikes in war time. We need that patriotic program too.'[56] At this point in his speech delegates shouted out: 'We need the Wagner Act. We think that is a patriotic policy too. Kirkland Lake.' In their view it was blatant hypocrisy to rely on this American example while refusing to consider an American type of collective-bargaining legislation.

These interruptions caused Mitchell to stray from his prepared address. He defended himself as a trade unionist by heritage and conviction and declared, 'I think you got a good agreement at Kirkland Lake – and don't you kid yourselves about that ... There is no justification for letting down the men who fight for us ... I do not care what arguments are advanced. I am not saying there is complete justice for the workers in all war plants ...' He then said that he had been told he was going to get a hostile reception and 'it takes courage to come down and talk to you and I said I would be glad to do it.' He reiterated that there was no complaint big enough to warrant ceasing one day in making war supplies, and expounded the Liberal government's solution to industrial problems: 'If there are grievances, there can be conciliation. Grown men can iron out their differences' and keep the industries running. He stuck to this position, which seemed irrelevant to what had actually occurred at Kirkland Lake – as the delegates at the convention realized.

Mosher thanked the minister of labour for attending the convention and indicated that although 'a few of our gathering may have felt it necessary to let off a little steam,' the majority appreciated 'the courage you have displayed in coming down here and speaking to this gathering.'[57] Despite these kind words, the response of the convention to Mitchell indicated more clearly than any resolution could do the complete alienation of the delegates from the government, its labour policy, and its labour department.

The Ontario labour minister, Peter Heenan, was given a much warmer welcome as he dropped a political 'bombshell' at the convention. He indicated that the Ontario government was preparing a collective-bargaining bill modelled on the Wagner Act to 'force those employers who are still living in the past to recognize their men, give them freedom of association, and then bargain with them collectively.'[58] He promised that the provincial government would present a bill next session; he asked labour representatives to help draft such a bill, and concluded that 'it will be a real workers' bill' with nothing insincere about it. The tone of his speech was entirely different from Mitchell's. Heenan declared that the workers had made their point: 'Collective bargaining will be legalized in the Province of Ontario.' He explained to the delegates (though it was the first anyone seemed to have heard of it) that there was a race between Ottawa and Ontario to see which government would be the first to enact an appropriate bill: 'But I don't care who takes the credit.' If Ottawa drafted a better bill, Ontario would copy it so long as 'it is the workers of the country who will benefit.'

Undoubtedly the existing rivalry between the Ontario and federal Liberal governments was present in this issue, but the real reason for this sudden change of attitude on the part of this previously anti-labour government was the political situation in Ontario. Throughout 1942 the CCF was attracting members, supporters, and revenue and becoming a credible alternative to the two old parties in the province: 'The greatest new source from which the CCF was deriving members and revenue was Ontario's mushrooming trade union movement.'[59]

Both nationally and in Ontario, the old parties began to respond to this dramatic change in the political climate. *Saturday Night* commented that 'the much tooted Heenan collective-bargaining bill is an obvious Ontario Liberal sop,' but after the convention both the Liberals and the communists (opposing the CCF's growing influence within the unions) tried to portray Heenan as an alternative and reliable friend of labour. Ultimately, Hepburn's resignation as premier in October 1942, and the resulting disintegration of the Ontario Liberal Party, made it impossible to continue this portrayal.

The other motivation behind the provincial government's promise of a new labour bill was the rising level of industrial unrest. Heenan told the CCL convention that the new bill was designed to resolve recognition disputes and would, therefore, reduce industrial conflict. Mosher was generous in his praise: 'It would be superfluous for me to express the thanks of this audience for your remarks. The applause has already indicated that.'[60]

Heenan's speech was a clever political move. He completely upstaged Mitchell, and disarmed the convention by expressing his government's good will and interest in labour's problems. While promising that his government would act by presenting a bill, he was still not very specific, but he had not proposed that workers give up their right to strike – nor had he suggested that the very existence of trade unionism and collective bargaining could be resolved by conciliation. In fact, the Ontario government was probably not as firm about its plans for the proposed bill as Heenan had indicated. However, Heenan had held out a ray of hope to the convention delegates, while Mitchell had merely repeated his government's intention to do nothing.

In 1942, the TLC convention was held three weeks prior to the CCL gathering and its delegates' reaction to the loss in Kirkland Lake was also one of deep concern. Several resolutions were passed in favour of a 'Wagner Act.' The resolution of the London Lodge of the Brotherhood of Railway Carmen declared that 'Kirkland Lake was a glaring example of the inability of the government to deal effectively with a situation of labour relations during a critical period.'[61] Many of the delegates wanted the TLC to hold a special convention to examine the current status of collective bargaining. Secretary-Treasurer D'Aoust advised against an open confrontation at a time when events were pushing the government towards the kind of legislation labour was seeking. The 'events' were the expansion of the labour movement, the escalation of industrial conflict, the growing disaffection of working people, and the consequent rise of support for the CCF. D'Aoust's reference was undoubtedly to the federal Cabinet's earlier promise (in February 1942) to proclaim an order enforcing PC 2685 in Crown corporations. Heenan's promise of an Ontario bill shortly thereafter demonstrated that there was some validity in D'Aoust's evaluation of the political situation. The TLC convention authorized its new executive to draft a model proposal, distribute it to its affiliates, and to lobby the government.

Thus the actions of the two congresses, while differing in detail, took the same general form. On the issue of legislative change the 'sober and responsible' leadership of the TLC was in general agreement with the 'militant and immoderate' leaders of the CCL. Although the direct relationship between the two congresses remained cool, after Kirkland Lake there was a greater union of purpose. For both congresses, the Kirkland Lake strike was a turning point. The bitter defeat was a reality that could not be obscured by mere declarations of principle or sympathetic speeches. The issues had been sharply defined. The problems could not be solved by the traditional methods of consultation, conciliation, and procrastination. For many people, the strike highlighted an obvious injustice that could only be remedied by legislation. For the government, the strike posed a dilemma. There was really no answer to the question, 'What about Kirkland Lake?' The issue could not be postponed or avoided because it was *inaction* that was under attack. The government had only two options – to repudiate the principles expressed in PC 2685 or to implement them. The latter option eventually became politically expedient.

POLITICAL RESPONSES

By the end of November, the prime minister was aware of the unity of organized labour and its anger at the delay in issuing the promised order concerning Crown corporations. He was disturbed at the inaction of his own ministers,[62] and he intervened to urge the immediate adoption of an order. This action resulted in the proclamation in December 1942 of PC 10802 (see appendix 1), which required Crown companies to operate under the principles of PC 2685. The order prevented the recurrence of another National Steel Car situation, but could not affect the Kirkland Lake situation or similar disputes in the 'private' sector. Nevertheless, PC 10802 made a uniform national labour code almost inevitable, since it made little sense to have the right to collective bargaining contingent upon the place where employees worked. Such an inconsistency could only inflame both labour and public opinion. PC 1003 was eventually issued some fourteen months later.

The political effects of the Kirkland Lake strike are difficult to assess, although it is clear that it was 'the last straw' and prompted the labour movement to concentrate on political action in order to bring about legislative change. Kirkland Lake demonstrated that the government was not only opposed to the Wagner Act principles, but also was prepared to use its influence to promote 'company unionism' and to undermine the right to strike. It seemed that the creation of a new political vehicle was the only answer. Thus, the CCL moved gradually towards a more formal alliance with the CCF. During the same period, the federal government modified the decision on collective bargaining that had been made at the height of the Kirkland Lake strike, following a consideration of the 'McLarty memo' (see chapter 8).

Just one day after the end of the strike two important by-elections were held. The most famous one involved the riding of South York where CCF candidate Joe Noseworthy, with strong labour support, defeated the newly drafted Conservative Party leader, Arthur Meighen.[63] The strength of the CCF grew after this by-election victory and increased efforts were made by the CCF to persuade individual unions to affiliate to the party, in the tradition of the British Labour Party.

From the point of view of organized labour, an equally important contest was the election of Humphrey Mitchell in Welland. At the time it seemed that the real victor was William Lyon Mackenzie King, as many editorials commented.[64] King was delighted.

What we have gained at one stroke was the greatest of all – two Ministers chosen from outside the House of Commons – each bringing new and much needed strength to the Ministry (great risk in the case of both) ... In the case of Mitchell, the best possible man for the Minister of Labour; but more than this – Meighen defeated and out of Parliament altogether. Not only Meighen defeated but the Leader of the Conservative Party defeated, and defeated in the strongest riding in Toronto, which means the strongest Tory riding in all of Canada.[65]

His hand-picked labour minister had been elected; his old enemy, Meighen, would not be sitting across the aisle from him, and Premier Hepburn, who had publicly supported both Meighen and Mitchell's opponent, had been discredited. Hepburn's appearance on the same platform with George Drew had had no substantial effect.

In the long run, however, the defeat of the Kirkland Lake miners and the government's refusal to change its labour policy benefitted the CCF. The CCF had decided to run a candidate in South York against Meighen in the absence of a Liberal candidate. The party also considered whether it should keep a candidate in the field against Mitchell. David Lewis, then national secretary of the CCF, had told Millard that the party decision 'will be greatly affected by what Mitchell does in the Kirkland Lake situation.'[66] His performance in that strike and the adamant demands by CCL leaders that he be opposed in the by-election settled the issue.[67] In the by-elections, labour and CCF organizations co-operated in an effort to defeat both Meighen and Mitchell. The Ontario CCF asked union members to man the polls in both by-elections.[68]

At the time of the by-elections, the CCF was considering a fund (publicized by its paper, *New Commonwealth*) to help the miners in Kirkland Lake to pay their debts. This initiative finally took the form of an appeal to its Ontario membership and to all other provincial councils in the party for *both* Kirkland Lake and the South York election.[69] This joint appeal is interesting in that it demonstrates the close relationship between the issues in Kirkland Lake and the political program adopted by the CCF. Labour and the CCF began to share common political objectives. In these circumstances, organizational co-operation of this kind seemed beneficial to both groups.[70]

Mitchell's election ensured there would be no immediate change in the government's labour policy. To labour, Mitchell had become synonymous with the company-union solution to the issue of recognition and collective bargaining. But the success of the CCF in South York indicated that times might be changing. Horowitz has written that this by-election 'was the signal for a veritable eruption of public support for the CCF.' The thousands of workers who had joined the new CCL unions were 'one of the main sources of the CCF's increased popular support. The CCL was now solidly established; pro-CCF sentiment among its members had risen dramatically and the national atmosphere had changed radically. The time for caution was passing.'[71] These developments did not bode well for the Liberals.

The main national issues in these two important by-elections had been conscription and post-war reconstruction, but the government's labour policy was also an issue in Welland where the CCF candidate had the support of local trade unionists and the Ontario CCF Trade Union Committee.[72] At one point in the campaign, when Hepburn and Drew were campaigning against Mitchell, Hepburn attacked King and his plebiscite on conscription. When he turned to attack his favourite bogey, John L. Lewis and the CIO, he was interrupted by a voice in the audience yelling, 'What about the mine owners?' Hepburn replied that he would speak of them later, but never returned to the subject.[73] In South York, labour dissatisfaction was also an undercurrent influencing some voters. The new CCF member, Joe Noseworthy, indicated such a trend on the night of his election. His victory, he said, was 'certainly the sign of new vigour for the CCF ... It is inescapable proof that the people of South York, who are representative of the people of Canada, want labour to have a say in the government and in our war effort.'[74]

The CCF realized that Ontario was the key to trade union support. The Ontario CCF executive met with leading trade unionists in the province and a CCF Trade Union Committee was formed. Among the members of this committee were Charles Millard of the steelworkers' union, Fred Dowling of the packinghouse workers' union, Sol Spivak of the amalgamated clothing workers, Hyman Langer of the ILGWU (International Ladies' Garment Workers' Union), and representatives from the typographical and printing pressmen's unions, as well as Max Federman of the Fur Workers' Union. All of these men had been prominent in the efforts to organize support for the Kirkland Lake miners.

In September 1942, Clarie Gillis, MP from Cape Breton South, a long time member of the United Mine Workers' Union and an important ally during the Kirkland Lake strike, accepted an invitation from this committee to come to Toronto to work on the affiliation of unions with the CCF. It was recognized that if the CCF would encourage the affiliation of unions in Ontario, 'the ball will roll automatically in other parts.'[75] Gillis was optimistic and enthusiastic. He had a deep faith in the need for the CCF to have its roots in the labour movement. He wrote to David Lewis at this time: 'I have already met and talked to most of the dignitaries in the union movement on the Congress side and the AFL side and to date I have not found any who disagree with the proposal to affiliate the unions. We are receiving active cooperation from both sides and I think before we are through here that we are going to have a pretty well balanced movement. The professors on the one side and the proletariat on the other.'[76]

For four months, Gillis toured the province, spoke to many local unions, and emphasized the importance of direct union affiliation to the CCF. He was a persuasive speaker and as an experienced union man was given serious consideration wherever he went. During that period, nineteen locals with a total of 12 000 members voted to affiliate with the CCF and pay per capita dues. The CCF Trade Union Committee sponsored an Ontario conference during this period that attracted delegates from sixty-nine local unions, most of them affiliated with the CCL. The conference endorsed the CCF as 'the political arm of labour,' a phrase which was to be heard with increasing frequency. Only eight months after Noseworthy's victory, the CCL convention, with only one dissenting vote, passed a resolution recognizing the CCF as labour's representative in Parliament, and recommended to its affiliated unions 'that they study the program of the CCF.'[77] 'During the year, the labour movement grew in numbers and the CCL's involvement in politics by supporting the CCF Party, became more evident. But despite mounting labour unrest, the Cabinet did not consider responding to the demands of the CCL and the TLC for a new national labour code modelled on the Wagner Act.'[78] Indeed, the long delay in implementing the promised policy with regard to Crown corporations further angered the labour movement, and was a major factor in diminishing labour support in 1943. By the fall of 1942, a dispute in the steel industry was becoming a test of the government's wage policy in the same way that the Kirkland Lake dispute had been a test of the government's policy on collective bargaining. The steel dispute was successfully deferred to 1943,[79] but in the process the government became incensed at what it considered Millard's and the steelworkers' belligerence. It was King's impression that the 'steelworkers and the CCF were playing politics.'[80] This, of course, was accurate, in that wages, working conditions, and collective-bargaining rights continued to be restricted by Cabinet orders emanating from a government perceived to be unresponsive to labour's interests, if not actually anti-labour. Only the CCF offered an alternative at this time. When combined with all the ill-will that had gone before, the wage policy dispute further widened the gap between labour and the government and pushed the labour movement into the arms of the CCF.

In subsequent months, labour's increasing support for the CCF and the rising popularity of that party produced a more sympathetic response from the old parties to labour's problems. The Ontario Liberals had promised a collective-bargaining bill. In April 1943, the Hepburn government enacted the Ontario Collective Bargaining Act (one year after Heenan's promise to the CCL), in an unsuccessful attempt to gain labour support in the 1943 Ontario election. The new act adopted the basic principles of the Wagner Act, including compulsory recognition and collective bargaining with the majority union and a specialized tribunal (the Labour Court) to enforce the legislation.

Not to be outdone, George Drew's Conservatives adopted a Twenty-Two Point program for the 1943 campaign that promised comprehensive economic measures and social security, 'advanced and fair labour laws,' and 'comprehensive collective-bargaining legislation.' The federal Conservatives also moved left when they chose Progressive Premier John Bracken of Manitoba as their new leader, and drafted a new program designed to combat the CCF. Toward the end of 1942 the marked rise in CCF support and its developing alliance with the labour movement gave the prime minister some concern, although his main interest was not yet with the CCF. When, in 1943, Mackenzie King did become much more disturbed by the rising strength of the CCF,[81] he established a parliamentary committee to design a social security program. In response to industrial unrest he ordered the new National War Labour Board to conduct a public inquiry into labour relations in Canada.

There were local political repercussions in Kirkland Lake after the strike. The community remained divided on both industrial and political issues. After the strike, Anne Shipley succeeded in displacing R.J. Carter as reeve. In 1943, Bob Carlin ran for the CCF in the provincial election and was elected as a member for Sudbury. He formed part of the thirty-four-member CCF Official Opposition. The success of the CCF in that historic election was the culmination of labour and party organizing that had begun in the South York and Welland by-elections. The part labour played in the growth of the CCF was indicated by the composition of the CCF caucus which, in 1943, included nineteen trade unionists. Carlin attributed his victory to his role in the Kirkland Lake strike. Miners and their wives, 'politicized' during the strike, worked hard for him in the 1943 campaign. Because his victory was assured, he himself campaigned for other CCF members, including Charlie Millard and Ted Jolliffe. During this election campaign the CCF raised the Kirkland Lake strike as an issue in some industrial centres, since throughout 1942 and 1943 Kirkland Lake had been the dispute most frequently mentioned by organized labour as an example of government's hostility to labour. Carlin visited Hamilton where the local CCF constituency organization had adopted the slogan 'Don't Forget Kirkland Lake.'[82] Carlin's personal assessment is that the Kirkland Lake strike and the government's labour policy 'led to the election of half of the 34 [CCF] members that found their way to the legislature.'[83]

The great gains made by the CCF in the Ontario election were attributed by the prime minister to labour's resentment of the government's wage stabilization policy. That was undoubtedly true, as the battle between organized labour and the government over the wage policy was continuing in the steel industry. Within the local campaigns, however, Kirkland Lake and the general issue of collective bargaining were still fresh in the minds of Ontario voters. Characteristically, King also blamed 'some of my colleagues [who] have become surrounded by interests that are ... not sympathetic to labour.' He was concerned by the collapse of the Liberal Party in Ontario and feared it might 'be the beginning of the end of the power of the Liberal Party federally.' This fear grew in his mind when, in August 1943, the federal Liberals lost four by-elections – two to the CCF. King believed this serious setback resulted from the 'bad handling of labour policies' and from poor party organization.[84] King became convinced that the alienation of labour was the greatest threat to the chances of a Liberal Party win in the next election. In August 1943, he visited the TLC convention, and in September he presented a new platform to the National Liberal Federation and made a strong appeal to the labour vote.

The results of the 1943 Ontario election set off a chain of events that induced the federal government to alter its labour policy. At its 1943 convention, the CCL endorsed the CCF as the political arm of labour. The TLC maintained its traditional policy of non-alignment, but only after a tumultuous debate on the floor of its convention. In any case, the TLC did become more active through its Political Action Committee, and this increased CCF support, although it did not result in an organic link. Both federations attacked the government's wage policy and criticized the tardiness of the NWLB in clarifying labour policy.

After the Ontario election, the Drew government continued to enforce the Ontario Collective Bargaining Act. The Labour Court mechanism was criticized by labour for its 'legalism' and formality, but organized labour generally supported the act. Despite its imperfections, it was serving a need. It effectively ended the need for recognition strikes. In its first six months of operation, the Labour Court was preoccupied with certification proceedings, and received 130 applications affecting approximately 80 000 persons.[85]

The Ontario act was also an important influence on the federal government. Just as the public hearings preceding the Ontario act had provided a public forum, which labour used to mobilize support for its position, the public hearings of the NWLB gave labour a national platform from which to air its grievances. At the end of 1943, both the majority and minority reports of the National War Labour Board recommended compulsory collective-bargaining legislation. By this time the federal government was prepared to enact it, particularly as it had decided to continue the wage policy that was so unpopular with the labour movement.[86]

The national labour code (PC 1003), guaranteeing the right of collective bargaining, was now viewed as a *political* necessity. This perception was in marked contrast to the view held by the government only two years before when, during the Kirkland Lake strike, the government had specifically decided not to intervene or enact a Canadian Wagner Act to resolve such recognition strikes. As organized workers sought a new status in industry and government, the political consensus the government wanted to maintain was crumbling. The country experienced unprecedented levels of industrial unrest in 1943, in accordance with Eugene Forsey's earlier prediction: 'King is never tired of proclaiming "national unity." It does not seem to have struck him that national unity can be broken vertically, as well as horizontally, and that by his present policy he is sowing the seeds of the bitterest class-struggle that would shatter national unity to fragments.'[87]

PC 1003 was not proclaimed until February 1944. The immediate political impact of the order was to undercut labour's opposition to the government, but because the legislation was implemented in the form of an order-in-council, it would be in effect only for the duration of the war. The fear that these legislative and organizational gains would be reversed after the war resulted in a new wave of industrial unrest designed to ensure that they would be preserved. Ultimately wartime advances were maintained in the post-war era, as in 1948 the Industrial Relations Investigation Act (IRDI Act) replaced PC 1003 and the IDI Act at the federal level and thereby entrenched the principles of compulsory collective bargaining and compulsory conciliation into our modern industrial relations system. The provinces either opted into this legislation or adopted similar acts of their own.

'REMEMBER KIRKLAND LAKE'

The Kirkland Lake strike was more than a struggle for recognition and collective bargaining rights for the 4000 miners who worked there; it was also part of a struggle to persuade the government that the rights of organized employees deserved legislative recognition. At Kirkland Lake the government adopted a position of 'non-intervention' that tacitly supported the mine owners. The miners were defeated because their new labour organization had neither economic nor political power. Their defeat convinced them that ultimate success would require both. 'Kirkland Lake' demonstrated to them, and to many others, that the 'justice' of their cause was less important than the power of their organizations. As Logan noted in 1948: '[the Kirkland Lake strike] drew the attention of a large public to the inequities of the situation and to the chagrin and resentment of the trade union world in not having them redressed. To many minds it appeared that the workers had struck for things that were reasonable and basic in a democratic system.'[88] Coates observed many years later that 'the party forming the government between 1935 and 1944 did not accept labour union demands for a change in national labour policy until labour achieved sufficient strength during a war emergency period to join with the CCF party and appear to threaten the survival of the Liberal party and the government.'[89]

'Remember Kirkland Lake' was to become an effective rallying cry in building a successful labour organization and an effective political party. At the 1942 CCL convention, Secretary-Treasurer Pat Conroy told the delegates: 'I assure the men of Kirkland Lake that they have made a contribution which will ring down through the years to come.'[90] The Canadian Unionist best conveyed the strike's legacy when it printed the following warning: 'Let the government note one thing – the workers of Canada will not give up their demand for freedom of association, for union recognition, for the right to bargain collectively with representatives of their free choice ... they will remember Kirkland Lake ... and they will keep on striving to establish their rights until success is achieved.'[91]

The Kirkland Lake strike was lost and therefore did not in itself resolve anything. The peak of labour/government conflict would occur in 1943. Nevertheless, the strike had far-reaching repercussions. Its loss dramatically highlighted the inadequacy of the government's labour policy and spurred the labour movement to become more politically active and unified in its demand for collective-bargaining legislation. The memory of this bitter conflict was an inspiration during the two-year struggle for a labour code that would acknowledge the existence of the new industrial unions. By 1943, a political situation was created in which the federal government, for a variety of reasons, felt impelled to try to pacify labour's industrial and political protests by enacting collective-bargaining legislation (PC 1003). Consequently, the central issue of union recognition, over which the Kirkland Lake strike was waged and lost, was ultimately won.

Appendix 1:
Wartime orders-in-council

PC 2685, 19 JUNE 1940

This order contained a declaration of principles for the orderly conduct of industrial relations during the war. It recommended that fair and reasonable standards of wages and working conditions should be observed; that there should be no undue extension of hours; and that proper precautions should be taken to ensure safe and healthful working conditions. The order recognized the right of workers to organize in trade unions and to bargain collectively, recommended that disputes should be settled by negotiation or with the assistance of conciliation services, and suggested that collective agreements should provide machinery for adjusting grievances.

PC 2686, 19 JUNE 1940

This order provided for a National Labour Supply Council to advise on any matters touching labour supply for industry that might be referred to it by the minister of labour. The council consisted of representatives of business and labour, but was abolished on 24 February 1942.

PC 5922, 25 OCTOBER 1940

This order created an Interdepartmental Committee on Labour Co-ordination of senior civil servants. Its functions were to devise means of meeting labour requirements, to co-ordinate the activities of federal agencies, and to secure the co-operation of the provinces. It was supposed to consult the National Labour Supply Council to obtain the views of business and labour and to secure their co-operation, but was not very successful in achieving this goal.

PC 7440, 16 DECEMBER 1940

This was the first comprehensive government policy designed to control wages. It covered all industries under the extended IDI Act, and served as a guide for boards created under that act. On 24 October 1941, it was repealed and replaced by PC 8253. This order, which was proclaimed at the same time as the government's comprehensive price policy, extended wage control to all employers and generalized the policy of PC 7440. It amounted to a wage freeze and the extension of cost-of-living bonuses. Provision was made for permanent enforcement machinery in the form of a National War Labour Board and nine regional war labour boards.

PC 4020, 6 JUNE 1941

This order established the Industrial Disputes Inquiry Commission (IDIC) to supplement the appointment of formal conciliation boards under the IDI Act. It was hoped that the IDIC would provide a speedy alternative for resolving disputes without the need of conciliation.

PC 4844, JULY 1941

This order permitted the IDIC commissioners to examine allegations of discrimination against workers for their trade union activity.

PC 7307, 16 SEPTEMBER 1941

This order inaugurated a system of compulsory strike votes. Strikes were prohibited until after a vote of the workers concerned. The minister of labour was to be informed of a proposed strike. At his discretion he might direct that a vote be conducted among those who in his opinion were affected by the dispute. A strike could then take place only after a majority of those entitled to vote were in favour of such action.

PC 10802, DECEMBER 1942

This order declared the right of employees of Crown corporations to join unions and bargain collectively.

PC 1003, 17 FEBRUARY 1944

This order declared the right of workers to join unions and to bargain collectively, established administrative machinery to assist that process, and prohibited certain practices of both employers and unions. The rights of labour, the new unfair labour practices, and the powers of the administrative board were all defined broadly.

Appendix 2:
The Mohawk Valley Formula

The 'Mohawk Valley Formula' played an important role in employers' anti-union offensive in the United States. It was a recognizable strategy, which was summarized by the National Labor Relations Board (NLRB) as follows:

First: when a strike was threatened, the employer labelled the union leaders as 'agitators' to discredit them with the public and their own members. He implied that the union members represented a small minority that was imposing its will upon the majority. At the same time, management would put out various types of propaganda 'falsely stating the issues involved in the strike so that the strikers appear to be making arbitrary demands, and the real issues such as the employer's refusal to bargain collectively, are obscured.' Concurrently with these moves the employer would threaten to move or shut down the plant and would thereby align the influential members of the community against the strike.

Second: when the strike was called, the employer's cry was for 'law and order' so that the community would amass legal and police weapons against the imagined violence.

Third: public sentiment was stimulated by a 'mass meeting,' which could aid the employer in exerting pressure on the authorities.

Fourth: the employer encouraged 'the formation of a large armed police force to intimidate the strikers and to exert a psychological effect upon the citizens.'

Fifth: in order to heighten the demoralizing effect of such measures – all designed to convince the strikers that their cause was hopeless – the employer or an employer-controlled association of 'loyal employees' sponsored a highly publicized 'back-to-work' movement. This strategy caused the public to believe that most of the employees desired to return to work and thereby won sympathy for the employer. The back-to-work movement also enabled the employer to operate the plant with strike-breakers, to gain information about the extent of the strikers' solidarity, and to continue to refuse to bargain collectively with the strikers.

Sixth: the employer worked to ensure the continuation of the show of police force. The presence of such a force demoralized the strikers, ensured that employees who returned to work would remain there, and tried to force the remaining strikers to capitulate. If necessary, the employer would 'turn the locality into a war-like camp through the declaration of a state of emergency tantamount to martial law and barricade it from the outside world so that nothing may interfere with the successful conclusion of the "Formula," thereby driving home to the union leaders the futility of further efforts to hold their ranks intact.'

This summary is based on David J. Saposs and Elizabeth T. Bliss, *Anti-Labor Activities in the United States* (New York: League for Industrial Democracy, June 1938), pp. 19–21.

Notes

PREFACE

1 H.A. Logan, *Trade Unions in Canada* (Toronto: Macmillan Co. of Canada Ltd 1948), p. 547
2 William Arnold Martin, 'A Study of Legislation Designed to Foster Industrial Peace in the Common Law Jurisdiction of Canada' PHD dissertation, University of Toronto 1954), p. 346

CHAPTER 1

1 Fred Dixon speaking in his own defence at the trial of the leaders of the Winnipeg General Strike, 1919 (*Textile Labor*, July–August 1973).
2 Philip Murray, quoted in Irving Bernstein, *The Turbulent Years* (Boston: Houghton Mifflin Company 1971), p. 145
3 Testimony of Louis D. Brandeis before the United States Commission on Industrial Relations, 1919, quoted in E.W. Bakke, C. Kerr, and C.N. Anrod, *Unions, Management and the Public* (New York: Harcourt, Brace and World, Inc. 1967), p. 244
4 *Strikes and Lockouts in Canada* (1941) tabulates lost days in this strike on the basis of 2800 persons in 1941 and 2000 persons in 1942, which is low. The number of miners who struck in November 1941 was approximately 3700.
5 *Ontario Sessional Papers* (1942 Report), p. 6
6 J.L. Granatstein, *Canada's War* (Toronto: Oxford University Press 1975), p. 285. The best discussion of the evolution of the welfare state is in chapter 7 of this book.
7 William Arnold Martin, 'A Study of Legislation Designed to Foster Industrial Peace in the Common Law Jurisdiction of Canada' (PHD dissertation, Univer-

sity of Toronto, 1954), p. 285. The following statistics indicate trade union membership in Canada.

Year	Trade union membership	Total non-agricultural force
1931	310 544	2 427 000
1932	283 096	2 209 000
1933	285 720	2 168 000
1934	281 274	2 403 000
1935	280 648	2 452 000

8 Canada, Department of Labour, *Report, Labour Organizations in Canada* (Ottawa: King's Printer 1946), p. 15

9 Irving Abella, 'The CIO, the Communist Party, and the Foundation of the CCL 1936–41,' CHAR (Canadian Historical Association Report) 1969

10 Department of Labour, *Report, Labour Organizations in Canada* (1949), p. 15

11 Canada, Department of Labour, *Earnings and Hours of Work in Manufacturing and Employment and Average Weekly Wages and Salaries 1939–46*; *Report, Labour Organizations in Canada* (1941), p. 6. There is a parallel between the level of industrial production during the war and trade union growth. Those industries that expanded the fastest were unionized the most quickly.

12 *Report, Labour Organizations in Canada* (1942), p. 41

13 *Ibid.* (1945), p. 16

14 Canada, Department of Labour, *Strikes and Lockouts in Canada* (Ottawa: 1941, 1942, 1943)

15 *Ibid.* (1941), p. 1; (1942), p. 1

16 *Ibid.* (1977); in 1946, which is often considered a peak year of industrial unrest in Canada, only one trade union member in six was involved in strike activity. The most recent comparable example of membership participation in strike activity was in 1976 on account of the political 'National Day of Protest' against wage controls.

17 *Strikes and Lockouts in Canada* (1942), p. 1

18 *Ibid.*

19 Martin, 'A Study of Legislation,' p. 29

20 Irving Bernstein, *The Turbulent Years* (Boston: Houghton Mifflin Co. 1971), p. 353

21 Daniel Coates, 'Organized Labour and Politics in Canada: The Development of a National Labour Code' (PHD dissertation, Cornell University 1973), p. 34

22 Martin, 'A Study of Legislation,' pp. 292, 307

23 Kenneth McNaught, *A Prophet in Politics* (Toronto: University of Toronto Press 1963), p. 293

24 J.L. Cohen, *Collective Bargaining in Canada* (Toronto: Steelworkers' Organizing Committee 1941), p. 18
25 Martin, 'A Study of Legislation,' pp. 299–302
26 *Ibid.*, p. 300. 'The cleavage among delegates with reference to this resolution was roughly between stronger, older craft groups and weaker, semi-industrial unionists. It should be noted that this issue was a part of the basic issue separating the craft and industrial, which resulted in the AF of L and CIO split.'
27 Irving Abella, *Nationalism, Communism and Canadian Labour* (Toronto: University of Toronto Press 1975), chapter 2. Canadian historians have emphasized AFL 'domination' in the TLC's execution of this purge. While this was one factor, the *internal* structural tensions between the older craft and newer industrial unions were important for the same reasons that they were in the United States.
28 *Report, Labour Organizations in Canada* (1939), p. 31
29 *TLC Convention Proceedings* (Ottawa 1942)
30 *Report, Labour Organizations in Canada* (1940), p. 23
31 *Ibid.*, p. 24
32 The CFL was conservative in the sense that increasingly it took management positions on such basic questions as collective bargaining, often because it opposed the new industrial unions' American origins.
33 *Report, Labour Organizations in Canada* (1940), p. 24
34 *Canadian Unionist*, July 1941
35 CCL executive board minutes, 1941–42, vol. 99, Canadian Labour Congress papers, Public Archives of Canada (hereafter cited as CLC and PAC respectively)
36 See table in n. 7 for organizational figures.
37 See table in n. 7 for an indication of the spectacular increase in union membership.
38 *Report, Labour Organizations in Canada* (1941), p. 15
39 *Ibid.* (1942), p. 21
40 *Ibid.* (1946), p. 14
41 Jack Williams, *The Story of Unions in Canada* (Toronto: J.M. Dent and Sons [Canada] Ltd, 1975), p. 182
42 Bernstein, *The Turbulent Years*, p. 769
43 *Ibid.*, pp. 61, 39
44 *Ibid.*, p. 685
45 *Ibid.*, p. 684
46 *Ibid.*, p. 773
47 *Ibid.*, pp. 775–6

48 *Ibid.*, p. 718; Richard Polenberg, *War and Society* (Philadelphia: J.B. Lippincott Company 1973), p. 7. For example, during the war, Sidney Hillman was appointed to the National Defense Advisory Commission by FDR and was a co-director of the Office of Production Management.

49 Reginald Whitaker, *The Government Party: Organizing and Financing the Liberal Party of Canada 1930–58* (Toronto: University of Toronto Press 1977), chapters 3 and 4. There was considerable business financing and support of the Liberal Party and the mining industry contributed an important share of party support at both the federal and provincial levels (pp. 123–4).

50 Bernstein, *The Turbulent Years*, p. 322

51 *Ibid.*, p. 787

52 *Ibid.*, pp. 330, 350

53 *Ibid.*, p. 653

54 H.A. Logan, *State Intervention and Assistance in Collective Bargaining* (Toronto: University of Toronto Press 1956), p. 14

55 T. Copp, 'The Impact of Wage and Price Control on Workers in Montreal 1939–47' (unpublished paper, May 1976), p. 2

56 *Ibid.*

57 Ruth Pierson, 'Women's Emancipation and the Recruitment of Women into the Canadian Labour Force in World War II' in S.M. Trofimenkoff and A. Prentice, *The Neglected Majority* (Toronto: McClelland and Stewart 1977), pp. 126–31

58 *Ibid.*, p. 127. Mothers of young children were encouraged to enter the labour force, despite the prevailing attitude that such a development was socially undesirable.

59 Order-in-council PC 6286, file 31, vol. 630, Canada, Department of Labour papers, PAC

60 Cohen, *Collective Bargaining in Canada*, p. 18

61 Pierson, 'Women's Emancipation,' pp. 125–30. In September 1942, the Ministry of Labour established an inventory of employable persons in Canada based on a registration campaign. This led to the adoption of a program for transferring mobile workers from areas of labour surplus to areas of short supply. To implement this and other policies, the Department of Labour and the director of NSS used the local offices of the National Employment Service of the Unemployment Insurance Commission. The NSS could thus match unemployed persons with war jobs or armed service requirements. The objective was to direct persons to employment in the order of importance to the war effort and to maintain essential civilian services.

62 Canada, Department of Labour, *Labour Gazette* (Ottawa 1943), p. 1613

63 *Northern Miner*, 13 November 1941. A mining spokesman estimated that 1500 miners had left. Management believed this loss contributed to unrest.

64 Charles Lipton, *The Trade Union Movement of Canada 1827–1959* (Montreal: Canadian Social Publications Ltd 1968), p. 267

65 Canada, Department of Labour, *Wages and Hours of Labour in Canada* (1940). I have used figures from this annual survey. The increase in wages in manufacturing in 1940 was 4.5 per cent and in the common factory category was 3.5 per cent for unskilled labour and 4.0 per cent for skilled and semi-skilled labour. In the metal trades, the average increase was 4.5 per cent. The Dominion Bureau of Statistics also conducted wage surveys (which I examined) that state there were not significant differences between the two surveys.

66 Department of Labour, *Wages and Hours of Labour in Canada* (1940–43). In 1941 there were considerable increases in wages in all industries averaging 10 per cent over 1940. The average increase between 1939 and 1941 was 15 per cent or approximately equal to the increase in the cost of living from August 1939 to 1941, which was 14.9 per cent. Between 1941 and 1942 the gradual upward trend in wages continued by 8.3 per cent even after the introduction of wage controls and cost-of-living bonuses.

67 *Ibid.* (1943). The Dominion Bureau of Statistics Survey, *Earnings and Hours of Work in Manufacturing* (November 1946), estimated that nearly two-thirds of this work force worked between forty and forty-eight hours. This did not include overtime, which workers did extensively during the war. There apparently was some reduction in the straight-time work week during the war years.

68 Order-in-council PC 7440 (16 December 1940)

69 CCL executive board minutes, 22 July 1941, vol. 99, CLC papers, PAC

70 *Ibid.*

71 Cohen, *Collective Bargaining in Canada*, p. 47

72 *Canadian Mining and Metallurgical Bulletin* XXXIII (July 1940): 296. One of the businessmen to join the government to aid the war effort was Mr G.C. Bateman, who was prominent in the mining industry; Howe appointed him as metals controller. Bateman, as the management nominee on the 1939 Teck-Hughes conciliation board, adamantly opposed the majority position in favour of union recognition, and in July 1941 warned his government colleagues that 'agitators from abroad' were making progress in Ontario's mines.

73 Coates, 'Organized Labor and Politics in Canada,' p. 77. The TLC convention criticized the labour policy and called for a Canadian Wagner Act. TLC leaders did not actively implement the resolution. They supported PC 2685 but called for rigorous enforcement of the principles, particularly in the government's relations with its own contractors (R. Bothwell and W. Kilbourn, *C.D. Howe* [Toronto: McClelland and Stewart 1979], p. 162).

74 Cohen, *Collective Bargaining in Canada*, p. 26

75 Coates, 'Organized Labor and Politics in Canada,' p. 91; Bernstein, *The Turbulent Years*, describes the American experience.

76 Coates, 'Organized Labor and Politics in Canada,' p. 65: 'No specific demands were made but the delegates claiming great employer resistance to acceptance of basic principles of union recognition and collective bargaining, were unified in calling for an order-in-council "setting forth the present rights of labour."'

77 *Ibid.*, p. 66

78 Bernstein, *The Turbulent Years*, p. 78

79 Cohen, *Collective Bargaining in Canada*, p. 33. J.L. Cohen was the foremost labour lawyer on the union side at this time in Canada. He had great rapport with the leading trade unionists of the day. He understood the aspirations of the 'new unionism' and its new leaders. Union leaders trusted him. They treated him both as a legal expert upon whom they could depend for legal advice, but also, in a different context, as one of their own who was invited to participate in organizational and political decisions taken by union leaders. For this reason the J.L. Cohen papers are an invaluable source for students of labour history in this period. From the 1937 Oshawa strike to the passage of PC 1003 in 1944, J.L. Cohen was involved in virtually every major industrial dispute of the CCL unions. By the end of the war, there were others like Ted Jolliffe and David Lewis who became competitors of Cohen's in the labour law field and to whom the trade unions began to turn with work.

80 Henry Ferns and Bernard Ostry, *The Age of Mackenzie King* (Toronto: James Lorimer and Co. 1976), chapter 7

81 Bradley Rudin, 'Mackenzie King and the Writing of Canada's Anti-Labour Laws,' *Canadian Dimension* (January 1972): 44. A more analytical and scholarly essay on development of King's ideas about trade unions is in Reginald Whitaker, 'The Liberal Corporatist Ideas of Mackenzie King,' *Labour/Le Travailleur*, vol. 2 (1977).

82 Rudin, 'Mackenzie King ...,' p. 46. Paul Craven, *'An Impartial Umpire': Industrial Relations and the Canadian State 1900–1911* (Toronto: University of Toronto Press 1980, pp. 247, 367)

83 Rudin, 'Mackenzie King ...,' p. 48

84 Martin, 'A Study of Legislation,' p. 256

85 Craven, *'An Impartial Umpire,'* chapters 6 to 11

86 Victor Levant, *Capital and Labour: Partners?* (Toronto: Steel Rail Educational Publishing 1977) pp. 65–7; Bruce Scott, '"A Place in the Sun": the Industrial Council at Massey-Harris 1919–1929,' *Labour/Le Travailleur*, vol. 1, 1976. During the 1920s, this industrial council movement was widespread, and Scott analyses it in a case study at Massey-Harris.

87 Cohen, *Collective Bargaining in Canada*, p. 14

88 *Ibid.*, p. 44; general correspondence, Interdepartmental Committee on Labour Co-Ordination, 17 July 1941, 28 August 1941, vol. 148, Lachelle files, Canada, Department of Labour. It is clear from the Department of Labour papers that once the government realized it could be construed that PC 7440 made PC 2685 compulsory, it quickly determined to correct that misconception.

89 *Ottawa Morning Journal*, 7 November 1941. This position was clearly stated by the minister of labour, Norman McLarty, in a speech on 7 November 1941.

90 Coates, 'Organized Labor and Politics in Canada,' pp. 71–2

91 *Ibid.*, p. 75

92 Copp, 'The Impact of Wage and Price Control,' pp. 5–6

93 Coates, 'Organized Labor and Politics in Canada,' p. 82

94 Cohen, *Collective Bargaining in Canada*, p. 40

95 Coates, 'Organized Labor and Politics in Canada,' pp. 85–6. See chapters 4 and 5.

96 *Ibid.*, p. 95

97 *Ibid.*, p. 96

98 William Lyon Mackenzie King (hereafter WLMK) typescript diary, 15 October 1941, vol. 89, King papers, PAC

99 When the wage control order was in fact referred to the NLSC, labour's request for special treatment for low-wage workers was rejected. This would tend to narrow the prevailing wage differentials between the low-wage and high-wage regions of the country. The government argued that the purpose of the policy was to control wages and restrain inflationary pressures – not remove existing inequities in the wage structure. Under PC 8253, the national board could raise wages judged to be *unduly* low. This became the issue in the Peck case.

100 Coates, 'Organized Labor and Politics in Canada,' p. 91

101 *Industrial Canada* (July 1941), p. 190

102 *Ibid.*, p. 91

103 Memo to the prime minister from SWOC, Local 2352, 3 July 1941, vol. 38, CLC papers, PAC

104 J.L. Cohen to Norman McLarty, 18 May 1941, vol. 38, CLC papers, PAC

105 Conciliation Board Report and Minority Report, vol. 38, CLC papers, PAC

106 Memo on Peck Rolling Mills, 21 May 1941, WLMK memoranda, vol 310, King papers, PAC

107 *Ibid.*

108 Copp, 'The Impact of Wage and Price Control,' p. 5

109 J.W. Pickersgill, *The Mackenzie King Record*, vol. I (Toronto: University of Toronto Press 1960), p. 229; Bothwell and Kilbourn, *C.D. Howe*, pp. 162–4

110 Pickersgill, *The Mackenzie King Record*, vol. I, p. 229

111 *Royal Commission to inquire into the events that occurred at Arvida, P.Q. in July 1941* (Ottawa: King's Printer 1941), p. 7
112 CCL memorandum, 27 February 1942, p. 2, private papers of Mary Sefton

CHAPTER 2

1 Clark Kerr, 'Peace and the Bargaining Environment,' in Allan Flanders, ed., *Collective Bargaining* (Middlesex, England: Penguin Books Ltd 1969), p. 136
2 *Ibid.*, p. 129. Also Rex Lucas, *Minetown, Milltown and Railtown* (Toronto: University of Toronto Press 1971), p. 327
3 S.A. Pain, *Three Miles of Gold* (Toronto: Ryerson Press 1960), p. vii
4 *Ibid.*, p. viii
5 *Ibid.*, pp. 19, 51
6 *Ibid.*, p. 34
7 T. Tait, 'Haileybury: The Early Years,' *Ontario History* (December 1963): 200
8 Pain, *Three Miles of Gold*, p. 66
9 *Ibid.*, p. 68
10 *Ibid.*, pp. 83–4
11 Dick Hunter, international representative, United Steel Workers of America, personal interview with the author, 13 August 1973
12 S.A. Pain, *The Way North* (Toronto: Ryerson Press 1964), p. 198
13 Pain, *Three Miles of Gold*, p. 75
14 *Ibid.*, p. 78
15 *Ibid.*, p. 87
16 Brian J. Young, 'C. George McCullagh and the Leadership League,' *Canadian Historical Review* (September 1966), pp. 202, 203
17 Daniel Coates, 'Organized Labor and Politics in Canada: The Development of a National Labor Code' (PHD dissertation, Cornell University 1973), p. 33
18 Irving Abella, ed., *On Strike* (Toronto: James Lewis and Samuel, Publishers 1974), p. 106
19 *Ibid.*, p. 113
20 Neil McKenty, *Mitch Hepburn* (McClelland and Stewart 1967), p. 88; Abella, *On Strike*, p. 105. The mining magnates included such men as George McCullagh, William Wright, the Timmins brothers, and J.P. Bickell of the MacIntyre-Porcupine mines among others. They were a group set apart from the older financial community in that they were newly rich, mostly self-made men, speculators, and very visible socially and politically. They were vocal supporters of the 'free enterprise system' and conservative. In 1936 Hepburn was attracted to them because of their fierce individualism, their confident optimism about the future, and their money – which was dispensed generously on their own enjoyment. He became close to George McCullagh.

McCullagh, in his own right had become a wealthy man, but he was aided by his business connections with William Wright, the discoverer and a principal stockholder in the Wright-Hargreaves and Teck-Hughes gold mines. The two men bought and managed the *Globe* in 1936. Hence this group of businessmen were powerful because of their business, political, and media connections. This symbiotic combination of interests affected the development of the mining industry as described in H.V. Nelles, *The Politics of Development* (Toronto: Macmillan Co. of Canada Ltd 1974), and assured the mining industry tax breaks and subsidization in many ways by the public purse. Such good fortune gave the mine owners a sense of power and confidence. In the area of labour relations, they believed they could 'bust' any union. The Kirkland Lake strike proved that with the help of political friends they were right, at least in the short run.

21 Memo to the prime minister 11 November 1941, file 2817, William Lyon Mackenzie King (WLMK) memoranda and notes, vol. 276, Public Archives of Canada (PAC)

22 *The Globe and Mail*, 8 January 1942, p. 49

23 Pain, *Three Miles of Gold*, p. 78

24 *Ibid.*, p. 80

25 *The Globe and Mail*, 8 January 1942, p. 47

26 *Canadian Mining and Metallurgical Bulletin* XXXIII (July 1940): 296

27 Nelles, *The Politics of Development*, p. 435

28 *Canadian Mining and Metallurgical Bulletin* XXXIII (January 1940): 19; *ibid.*, XXXIII (June 1941)

29 R. Warren James, *Wartime Economic Co-operation* (Toronto: Ryerson Press, 1949); and J.W. Pickersgill, ed., *The Mackenzie King Record* (Toronto: University of Toronto Press 1960), vol. I

30 Memo to the prime minister, 11 November 1941, file 2817, WLMK memoranda, vol. 276, PAC

31 *Winnipeg Free Press*, 22 January 1942, Lawrence Jacks 'Wartime Gold Mining'

32 Memo to prime minister, 11 November 1941, file 2817, WLMK memoranda, vol. 276, PAC

33 *Toronto Star*, 1 November 1941

34 Memo to prime minister, 11 November 1941, file 2817, WLMK memoranda, vol. 276, PAC

35 *The Globe and Mail*, 8 January 1942, p. 49

36 Nelles, *The Politics of Development*, p. 109; e.g. Pain, *Three Miles of Gold* and *The Way North*; D.M. LeBourdais, *Metals and Men: The Story of Canadian Mining* (Toronto: McClelland and Stewart 1957)

37 Pain, *The Way North*, p. 187

38 *Ibid.*, p. 188; my emphasis

39 *Ibid.*, pp. 189, 191
40 Nelles, *Politics of Development*, p. 436. Nelles says, for example, 'much of the doctrinaire laisser-faire outlook of the mining industry was grounded in the strident self-confidence and individualism that universally accompanied the growth of the industry. To the confidence of new wealth was added the arrogance of success amid widespread failure' (in the 1930s).
41 Allan Fox, 'Management's Frame of Reference' in Flanders, ed., *Collective Bargaining*, p. 391
42 *Ibid.*, p. 393
43 'The Labour Situation at Kirkland Lake Ontario: An Important Statement,' p. 2, vol. 196, Canadian Labour Congress (CLC) papers, PAC
44 *Ottawa Morning Journal*, 19 November 1941
45 'The Labour Situation at Kirkland Lake,' p. 6, vol. 196, CLC papers, PAC. Often the international affiliation of the local Mine Mill and the indirect affiliation to the CIO was raised in this connection to prove that the union was an 'outside organization.' Given the mine-owners' perception, any union would have been a challenge to their 'managerial prerogative.'
46 *Ibid.*, pp. 1, 8
47 *Ibid.*, p. 9
48 *Ibid.*, pp. 1, 11
49 *Ibid.*, p. 6
50 C. Kerr and A. Siegel, 'Interindustry Propensity to Strike,' in Flanders, ed., *Collective Bargaining*, p. 147
51 Report of the 'Temiskaming Presbytery – United Church of Canada – Re: Kirkland Lake Labour Situation,' p. 2, box 1, J.R. Mutchmor general correspondence, United Church Archives
52 *Canadian Mining and Metallurgical Bulletin* XXXIII (September 1942): 421 and (November 1942): 511
53 Arnold Peters, MP Temiskaming, personal interview with the author at Kirkland Lake, 13 August 1973
54 Kerr and Siegel, 'Interindustry Propensity,' pp. 143, 147
55 Peters, personal interview with the author at Kirkland Lake, 13 August 1973
56 Pain, *Three Miles of Gold*, p. 83
57 'As the Collector Sees It,' unknown author, a collector and credit manager at Kirkland Lake, file 2780, J.L. Cohen papers, PAC
58 Union brief to the conciliation board regarding the industrial dispute between Teck-Hughes Gold mine and Local 240, p. 9, file 2780, J.L. Cohen papers, PAC
59 Union brief to the conciliation board, October 1941, p. 15, file 2879, J.L. Cohen papers, PAC
60 'As the Collector Sees It,' file 2780, Cohen papers

61 *Ibid.*

62 'The Kirkland Lake Labour Dispute,' vol. 196, CLC papers, PAC

63 *Census* (Canada) 1931, vol. 2, p. 427

64 File 209, vol. 638, Canada, Department of Labour, PAC. The ethnic make-up of the work force figures as of 30 November 1940 were the following:

Mine	*British or Canadian born* (per cent)	*Foreign born* (per cent)
Lake Shore	78	22
Wright-Hargreaves	80	20
Teck-Hughes	64	36
Sylvanite	58	42
Macassa	74	26
K.L. Gold	64	36
Toburn	60	40
Golden Gate	89	11
Bidgood	72	28
Upper Canada	87	13
	—	—
Average (also includes Larder Lake mines not listed here)	73	27

65 *Ibid.* Of the foreign-born work force in the mines, the proportion of naturalized workers to not naturalized was as follows:

Mine	*Naturalized* (per cent)	*Not naturalized* (per cent)
Lake Shore	62	30
Wright-Hargreaves	55	45
Teck-Hughes	58	42
Sylvanite	46	54
Macassa	48	52
K.L. Gold	44	56
Toburn	58	42
Golden Gate	56	44
Bidgood	22	78
Upper Canada	30	70
	—	—
Average (also includes Larder Lake mines not listed)	52	48

66 'Kirkland Lake Bulletin,' 2 January 1942, vol. 196, CLC papers, PAC

67 Eamon Park, SWOC organizer, radio talk, 28 January 1942, p. 2, box 2, J.N. 'Pat' Kelly papers, McMaster University Archives

68 Peters, personal interview with the author at Kirkland Lake, 13 August 1973

69 *Sudbury Star*, 3 March 1942

70 For example, John L. Lewis, president of the UMWA and the CIO; Philip Murray, president of the SWOC and the CIO; Silby Barrett, director of the Canadian CIO Committee

71 Wayne Roberts, ed., *Miner's Life: Bob Miner and Union Organizing in Timmins, Kirkland Lake and Sudbury* (Hamilton, MacMaster University 1979). Bob Miner makes a similar comment about the workers at Dome Mines in Timmins: 'Actually, once they lost a little of their fear, they found out that they resented the paternalism of the company as much as anyone. Nobody likes to be patronized. Nobody wants to be patted on the head instead of paid' (p. 2).

72 John B. Lang, 'A Lion in a Den of Daniels – A History of the IUMMSW in Sudbury, Ontario 1942–62' (MA dissertation, University of Guelph 1970), p. 12

73 Vernon Jensen, *Nonferrous Metals Industry Unionism* (New York: Cornell University 1954), p. 5. I have cited this book several times and I found it very interesting. However, it should be noted that all the Mine Mill people whose opinion of the book I solicited had reservations about it, including Orville Larson and Tom McGuire, who were active in the international union at the time. McGuire did not entirely disagree with Jensen, but he believed there were inaccuracies in the book. More importantly, he said, 'In my opinion [the book] does not truly reflect the character, purposes, objectives, and results of the IUMMSW' (McGuire, 1 October 1976, taped letter to the author).

74 Lang, 'A Lion in a Den of Daniels,' p. 13

75 Jensen, *Nonferrous Metals Industry Unionism*, p. 1

76 Martin Robin, *Radical Politics and Canadian Labour* (Kingston: Industrial Relations Centre, Queen's University 1968), p. 47

77 Lang, 'A Lion in a Den of Daniels,' p. 22

78 Jensen, *Nonferrous Metals Industry Unionism*, p. 3

79 *Ibid.* Also Patrick Renshaw, *The Wobblies* (New York: Doubleday & Co. Inc. 1967), p. 38

80 Lang, 'A Lion in a Den of Daniels,' p. 8

81 Paul Craven, *'An Impartial Umpire': Industrial Relations and the Canadian State, 1900–1911* (Toronto: University of Toronto Press 1980), pp. 246–52

82 Brian Hogan, *Cobalt: Year of the Strike, 1919* (Cobalt: Highway Book Shop 1978)

83 *Ibid.*, pp. 53–61. Also W. Roberts, ed., *Miner's Life* ..., p. 1. Management control of workers' lives was an issue in Timmins at this time.

84 Jensen, *Nonferrous Metals Industry Unionism*, p. 2. The name of the union changed from Western Federation of Miners in 1916.

85 Hogan, *Cobalt*, p. 81

86 Bob Carlin, financial secretary Local 240, transcript of taped interview with the author at Gowganda, August 1973, p. 6

87 *Ibid.*, p. 5

88 Hogan, *Cobalt*, p. 153

89 Carlin, transcript, p. 5

90 Hogan, *Cobalt*, p. 144

91 Carlin, transcript, p. 6

92 Lang, 'A Lion in a Den of Daniels,' p. 28; David Bercuson, *Fools and Wise Men: The Rise and Fall of the One Big Union* (Toronto: McGraw-Hill Ryerson Ltd 1978)

93 Carlin, transcript, p. 9

94 David Bercuson, 'Western Labour Radicalism and the One Big Union: Myths and Realities' in S.M. Trofimenkoff, ed., *The Twenties in Western Canada* (Ottawa 1972)

95 Lang, 'A Lion in a Den of Daniels,' p. 29

96 Tom McGuire, international representative IUMMSW, 1 October 1976, taped letter to the author

97 Jensen, *Nonferrous Metals Industry Unionism*, p. 5

98 Carlin, transcript, p. 12

99 Jensen, *Nonferrous Metals Industry Unionism*, pp. 35, 51

100 Lang, 'A Lion in a Den of Daniels,' pp. 30–1

101 Carlin, transcript, p. 14

102 Undercover man's report on Sudbury, 10 June 1937, box 267, Hepburn papers, 1937, Archives of Ontario

103 *Ibid.*

104 Union executive meeting, Kirkland Lake, 21 November 1937, report of an undercover man of the Ontario Mining Association, 26 November 1937, box 267, Hepburn papers, 1937

105 Union executive meeting, Kirkland Lake, 24 November 1937, report of an undercover man, 29 November 1937, *ibid.*

106 Lang, 'A Lion in a Den of Daniels,' pp. 31–3

107 Irving Abella, *Nationalism, Communism and Canadian Labour* (Toronto: University of Toronto Press 1973), p. 86. Confidential report of an undercover man on Kirkland Lake sent to the OPP, received 10 June 1937, box 267, Hepburn Papers, 1937. The chief organizers at Kirkland Lake were Thomas Church, J. Adler, George Zapotocki, A. Szyszko, W. Plaszkiewicz, A. Beherski, P. Vejacicp, Matewejczuk, M. Koziur; J. Uhryn, J. Rohatiuk, M. Zajac, J. Csonos, J. Poloski, M. Mokrij, J. Latkos, N. Robac.

108 Carlin, transcript, p. 15. Also Roberts, ed., *Miner's Life*, p. 5. The communists were active in the Timmins local, too.
109 Confidential report of an undercover man on Kirkland Lake and Timmins sent to the OPP, received 10 June 1937, box 267, Hepburn papers, 1937
110 *Ibid.*
111 Union executive meeting, Kirkland Lake, 21 November 1937 report of an undercover agent of the Ontario Mining Association, 26 November 1937, box 267, Hepburn papers, 1937
112 Carlin, transcript, p. 15

CHAPTER 3

1 M.J. Moky, Kirkland Lake miner, 19 October 1939, letter to J.L. Cohen, file 2762, J.L. Cohen papers, Public Archives of Canada (PAC)
2 J.L. Cohen, 15 November 1939, letter to Tom Church, file 2762, *ibid.*
3 M.J. Moky, 6 November 1939, letter to J.L. Cohen, file 2703, *ibid.*
4 J.L. Cohen, 29 November 1939, letter to Tom Church, file 2762, *ibid.*; and *Labour Gazette* (Ottawa 1940), p. 645
5 J.G. MacMillan, 7 December 1939, letter to board of conciliation, file 2762, *ibid.*
6 *Labour Gazette* (Ottawa 1940), p. 645
7 *Ibid.*, p. 198
8 *Canadian Mining and Metallurgical Bulletin* XXXIII (July 1941). G.C. Bateman was prominent in the mining industry and in June 1941 he was appointed metals controller by C.D. Howe.
9 *Labour Gazette* (Ottawa 1940), p. 647
10 *Ibid*, p. 648. The following quotations of the board in the text are all from the same source.
11 *Ibid.*, p. 649
12 Discussed in chapter 1. Between 1937 and 1939 most provinces enacted some protective legislation against dismissal for union activity.
13 *Labour Gazette* (Ottawa 1940), p. 651
14 *Ibid.*, p. 653
15 Union brief to the conciliation board regarding the industrial dispute at Teck-Hughes mine, 1939, p. 11, file 2780, J.L. Cohen papers, PAC
16 George Grube, 'The Strike at Kirkland Lake,' *Canadian Forum* (January 1942): 299
17 John B. Lang, 'A Lion in a Den of Daniels: A History of the IUMMSW in Sudbury, Ontario 1942–1962' (MA dissertation, University of Guelph 1973), p. 33

18 Grube, 'The Strike at Kirkland Lake,' p. 300
19 Bob Carlin, financial secretary, Local 240, transcript of taped interview with the author at Gowganda, August 1973, p. 23
20 Vernon Jensen, *Nonferrous Metals Industry Unionism* (New York: Cornell University 1954), p. 51. Reid Robinson was elected president of the IUMMSW in 1936 as a new unionist, who nevertheless had fought the communists in the Butte local and who seemed to follow his father Jim's philosophy of practical business unionism. Shortly after his handling of this dispute in Kirkland Lake in his old style, Robinson apparently changed his point of view. Even by 1941 he was recognized as the leader of the left faction, which included communists and communist sympathizers in the union.
21 *Ibid.*, p. 98. Irving Abella, *Nationalism, Communism and Canadian Labour* (Toronto: University of Toronto Press 1973), p. 87
22 Carlin, transcript, p. 24
23 Tom McGuire, international representative, IUMMSW, tape to the author, 1 October 1976
24 *Ibid.*
25 Carlin, transcript, p. 25
26 McGuire, tape 1 October 1976
27 Carlin, transcript, p. 25
28 *Ibid.*; it should be noted that at the level of the international union executive board, from 1942 on, Carlin was a consistent supporter of Reid Robinson and his left-wing group in an increasingly bitter dispute over who would control Mine Mill (Jensen, *Nonferrous Metals Industry Unionism*, p. 98). Carlin was elected CCF member of the provincial parliament in 1943. His continued close association with the communists in his union ultimately led to his eventual expulsion from the CCF in 1948 (Gad Horowitz, *Canadian Labour in Politics* [Toronto: University of Toronto Press 1968], p. 131f).
29 *Canadian Unionist* (October 1941), p. 104
30 Tom McGuire, 29 March 1941, 'Report of Kirkland Lake and Timmins Districts,' vol. 99, Canadian Labour Congress (CLC) papers, PAC
31 McGuire, tape, 1 October 1976
32 *Canadian Unionist* (October 1940), p. 127
33 *Ibid.* (May 1941), p. 302
34 McGuire, 'Report of Kirkland Lake and Timmins Districts'
35 *Canadian Unionist* (November 1940), p. 155
36 McGuire, 29 March 1941, letter to A.R. Mosher, vol. 99, CLC papers, PAC
37 Reid Robinson, 9 May 1941, letter to Norman Dowd, *ibid.*
38 Norman Dowd, 2 April 1941, letter to John L. Lewis, *ibid.*

39 A.R. Mosher, 18 January 1941, letter to Reid Robinson; and Norman Dowd, 4 April 1941, letter to Tom McGuire, *ibid.* The congress policy was to give occasional donations from its special organizing fund to help the organizing campaigns of its affiliates but *not* to *finance* such organizing campaigns. Such extensive financing was the responsibility of the national or international union. The congress did not have the funds for financial aid and organizing as such was not seen by many of its affiliates as its proper function. The CCL had total responsibility for organizing only in the case of chartered locals.

40 Norman Dowd, 4 April 1941, letter to Tom McGuire, *ibid.*; Dowd, 2 April 1941, letter to John L. Lewis, *ibid.*

41 Jensen, *Nonferrous Metals Industry Unionism*, pp. 26–9

42 John L. Lewis, 9 April 1941, letter to Norman Dowd, vol. 99, CLC papers, PAC

43 Reid Robinson, 9 May 1941, letter to Norman Dowd, *ibid.*

44 *Canadian Unionist* (April 1941), p. 273

45 W.H. Springer to attorney general, 28 August 1941; Agent No. 202 to H.S. McCready (deputy police commissioner) 3 September 1941, file 11.8 Provincial Police (Strikes 1942), Ontario Attorney General's papers, Archives of Ontario

46 Carlin, 10 June 1941, letter to A.L. Bloomfield, file 2879, J.L. Cohen papers, PAC

47 It is difficult to determine the membership support at this time. The CCL figures were overly optimistic, but by the fall of 1941 the union was confident that it had close to 90 per cent support.

CHAPTER 4

1 N. McLarty, 3 December 1941, letter to W.L.M. King, file 49, vol. 310 primary correspondence, King papers, Public Archives of Canada (PAC)

2 J.H. Brown, 14 July 1941, letter to Bob Carlin, file 2879, J.L. Cohen papers, PAC

3 Bob Carlin, 21 July 1941, letter to N. McLarty, *ibid.*

4 *Brantford Expositor*, 18 July 1941

5 Tom McGuire, 24 July 1941, letter to Reid Robinson, file 2879, J.L. Cohen papers, PAC

6 J.L. Cohen, *Collective Bargaining in Canada* (Toronto: Steelworkers' Organizing Committee 1941), p. 27

7 The National Labour Supply Council was established by PC 2686 on 19 June 1941 to advise the government on any matters affecting labour supply that

might be referred to it by the minister of labour. While it was merely an advisory body, it was the one wartime policy-making body in which labour had equal representation with business. It was established partly as a response to labour's criticism that the labour movement was not represented on bodies making war policies that affected its members.

8 Daniel Coates, 'Organized Labor and Politics in Canada: The Development of a National Labor Code' (PHD dissertation, Cornell University 1973), p. 84

9 H.D. Woods, *Labour Policy in Canada* (Toronto: Macmillan of Canada 1973), p. 51

10 Cohen, *Collective Bargaining*, p. 51

11 Order-in-council PC 4020 (6 June 1941). The preamble of the order notes that the extension of the IDI Act increased the number of applications for conciliation, and that 'a number may have reference to disputes which *prima facie* [do not] warrant the appointment of a board of conciliation.' In these circumstances, an IDIC commissioner could be appointed to 'carry out a preliminary investigation promptly and if a mutually satisfactory adjustment is not arrived at ... advise the Minister of the matters at issue *and whether the circumstances warrant the appointment of a conciliation board.*' The IDIC could be appointed 'whether or not conciliation has been applied for.'

12 McGuire, 3 August 1941, letter to David Lewis, national secretary, CCF, vol. 188, Co-operative Commonwealth Federation (CCF) papers, PAC

13 Cohen, *Collective Bargaining*, p. 53

14 William Arnold Martin, 'A Study of Legislation Designed to Foster Industrial Peace in the Common Law Jurisdiction of Canada' PHD dissertation, University of Toronto 1954), p. 256

15 Humphrey Mitchell was first elected to Parliament as a Labour member (with Liberal support) in 1930 from the constituency of Hamilton East. He won what had been a traditionally Conservative riding when at Mackenzie King's urging, the Liberals did not run a candidate. Mitchell won a strong victory over the Conservative candidate. This development was beneficial to the Liberal Party. The riding has since been a Liberal stronghold, where the party has drawn considerable support from the unionized working-class electorate. The Liberal Party thus gained a labour base in a working-class constituency that might otherwise have gone to the CCF. Mitchell was given several positions on boards and commissions dealing with labour relations during World War II on the basis that he was 'a labour man' and was appointed minister of labour in the Liberal government in 1941 to replace corporation lawyer Norman McLarty. Mitchell retained his portfolio until his death in 1950.

16 Rev. W. Cullis, 19 December 1941, letter to Dr J. Coburn, box 1, J.R. Mutchmor general correspondence, United Church Archives

17 *Northern News*, 8 August 1941
18 'Report of the Industrial Disputes Inquiry Commission on the Dispute Between Various Gold Mining Companies, Kirkland Lake, Ontario, and Certain of their Employees,' p. 2, file 209, vol. 638, Canada, Department of Labour, PAC
19 Cohen, *Collective Bargaining*, pp. 57–8
20 *Ibid.*
21 W.M. Thompson, 20 August 1941, letter to Hon. J.L. Ilsley, minister of finance, file 2879, J.L. Cohen papers, PAC. As of 6 August 1941, only four mines definitely agreed to adjust the cost-of-living bonus immediately, or as soon as a new index figure was available. The Morris-Kirkland mine, for example, which had gone into bankruptcy, applied to the government for exemption from the payment of the bonus.
22 Miners' Union Negotiating Committee recommendations on the IDIC Report, 7 August 1941, vol. 187, national CCF papers, PAC
23 Cohen, *Collective Bargaining*, p. 54
24 'Report of the IDIC,' p. 4, file 209, vol. 638, Canada, Department of Labour, PAC
25 *Ibid.*
26 *North Bay Nugget*, 12 August 1941
27 Cohen, *Collective Bargaining*, p. 54
28 H. Ferns and B. Ostry, *The Age of Mackenzie King* (Toronto: James Lorimer and Co. 1976), p. 207
29 Cohen, *Collective Bargaining*, p. 59
30 Brian Hogan, *Cobalt: Year of the Strike 1919* (Cobalt: Highway Book Shop 1978). In Cobalt, prior to World War I, the union existed and was active but was ignored by the companies.
31 'Lake Shore Mines Ltd. Workmen's Council,' file 2900, J.L. Cohen papers, PAC
32 Sometimes, employees' committees began spontaneously where there was no unionism. They were weak. Usually they became part of a union in time, as, for example, the Algoma local, which joined the United Steelworkers of America in 1940.
33 Cohen, *Collective Bargaining*, p. 58
34 Humphrey Mitchell, 23 October 1941 to J. Corbett, vol. 312, primary correspondence, King papers, PAC
35 *Ibid.*
36 *Ibid.*
37 Hogan, *Cobalt*, p. 1
38 George Grube, 'The Strike at Kirkland Lake,' *Canadian Forum* (January 1942): 299

39 M.W. Hotchkin, 30 August 1941, letter to registrar, board of conciliation, file 2879, J.L. Cohen papers, PAC
40 The 'holiday' tactic was simply a euphemism for a strike. Legally the miners could not strike until after the release of the conciliation board report. As the government sought to delay conciliation against the union's wishes, the one-day 'holiday' was the only way the miners could convey their frustration at the delay, and indicate the degree of support for the union.
41 *Northern News*, 15 August 1941
42 *Northern News*, *Toronto Star*, *Toronto Telegram*, 15 August 1941
43 *Northern News*, 15 August 1941
44 M.W. Hotchkin, 30 August 1941, letter to registrar, board of conciliation, file 2879, J.L. Cohen papers, PAC
45 N. McLarty, 3 December 1941 to William Lyon Mackenzie King (WLMK), file 49, vol. 310, primary correspondence, King papers, PAC
46 *Northern News*, 19 August 1941
47 Grube, 'The Strike at Kirkland Lake,' p. 300
48 *Northern News*, 26 August 1941
49 *Ibid.*, 29 August 1941

CHAPTER 5

1 N. McLarty, 3 December 1941, letter to William Lyon Mackenzie King (WLMK), file 49, vol. 310, primary correspondence, King papers, Public Archives of Canada (PAC)
2 McTague was on many conciliation boards at the time. He became chairman of the reconstituted National War Labour Board (NWLB) in 1943. Norman McLarty was federal minister of labour until he was replaced by Humphrey Mitchell.
3 Wilkinson was a prominent company lawyer.
4 J.L. Cohen was the foremost labour lawyer on the union side at this time in Canada. See chapter 1, n. 79.
5 *Northern News*, 2 September 1941
6 File 2894, J.L. Cohen papers, PAC. The union took preliminary steps to initiate a libel suit against the *Northern News*.
7 File 2879, J.L. Cohen papers, PAC. The Cohen papers contain all the briefs of the mining companies to the conciliation board except the Sylvanite brief.
8 Teck-Hughes brief, file 2879, J.L. Cohen papers, PAC
9 Bidgood brief, *ibid.*
10 Upper Canada Gold mine brief, *ibid.*
11 Kirkland Lake Gold mine brief, *ibid.*

12 Macassa brief, *ibid.*
13 Written comments following the presentation of the briefs, *ibid.*
14 Union brief, pp. 53–4; Lake Shore mine brief and Bidgood mine brief, file 2879, J.L. Cohen papers, PAC
15 Upper Canada mine brief, *ibid.*
16 Lake Shore mine brief, *ibid.*
17 Union brief, pp. 1–87, *ibid.*
18 Vernon Jensen, *Nonferrous Metals Industry Unionism* (New York: Cornell University 1954), pp. 53, 95
19 Union brief, p. 8, file 2879, J.L. Cohen papers, PAC
20 *Ibid.*, pp. 11–12
21 CCL executive meeting, 22 October 1941, vol. 99, Canadian Labour Congress (CLC) papers, PAC
22 Union brief, p. 17, file 2879, J.L. Cohen papers, PAC
23 *Ibid.*, pp. 18, 20
24 *Ibid.*, p. 22
25 *Ibid.*, p. 32a
26 *Ibid.*, p. 60
27 *Ibid.*, p. 46
28 Dick Hunter, international representative, United Steelworkers of America, personal interview with the author, 13 August 1973
29 Union brief, pp. 47–9, file 2879, J.L. Cohen papers, PAC
30 *Ibid.*, p. 50
31 *Ibid.*, p. 54
32 *Ibid.*, p. 56
33 *Ibid.*, p. 57
34 Bob Carlin, financial secretary, Local 240, transcript of taped interview with the author at Gowganda, August 1973, pp. 11, 37, 38
35 Union brief, p. 63, file 2879, J.L. Cohen papers, PAC
36 Notes of J.L. Cohen, 6 October 1941, *ibid.*
37 *Toronto Star* clipping, 6 October 1941, *ibid.*
38 N. McLarty, 3 December 1941, letter to WLMK, file 49, vol. 310, primary correspondence, King papers, PAC
39 'The Labour Situation at Kirkland Lake, Ont.: An Important Statement,' vol. 196, CLC papers, PAC
40 E. Long, 30 October 1941, letter to J.R. Mutchmor, J.R. Mutchmor personal correspondence, United Church Archives
41 George Grube, 'The Strike at Kirkland Lake,' *Canadian Forum* (January 1942): 300
42 Conciliation board report in *Canadian Unionist* (October 1941), p. 108

43 *Ibid.*
44 CCL executive meeting, 22 October 1941, vol. 99, CLC papers, PAC
45 *Toronto Star*, 9 December 1941
46 *Winnipeg Free Press*, 22 January 1942; see also chapter 2
47 Reid Robinson, 28 October 1941, letter to J.L. Cohen, file 2900, J.L. Cohen papers, PAC
48 *The Globe and Mail*, 24 October 1941
49 Officer McDougall's 13th report, 31 October 1941, file 11.8, Provincial Police (Strikes 1942), Ontario Attorney General's papers, Archives of Ontario
50 Long, 30 October 1941, letter to J.R. Mutchmor, J.R. Mutchmor personal correspondence 1941, United Church Archives
51 Cohen, 31 October 1941, letter to Reid Robinson, file 2900, J.L. Cohen papers, PAC
52 Long, 30 October 1941, letter to J.R. Mutchmor, J.R. Mutchmor personal correspondence 1941, United Church Archives
53 File 2900, J.L. Cohen papers, PAC
54 Resolution of Local 240, 27 September 1941, to WLMK, vol. 152, Prime Minister's office (PMO) files, King papers, PAC. A typical protest against PC 7307 contained in this file, came from Local 222, UAW at the Ford Motor Company. The union complained that the order allowed the Labour Minister to impose restrictions which would 'virtually abolish labour's right to strike.' In its view the order placed the 'Federal Department of Labour in a position of assisting anti-labour employers in their practice of coercion and intimidation of employees.'
55 See, for example, Eugene Forsey, 'Mr. King, Parliament, the Constitution and Labour Policy,' *Canadian Forum* (January 1942): 296
56 Order-in-council PC 7307, 16 September 1941
57 Quoted in G. Grube, 'Labour Law by Order-in-Council,' *Canadian Forum* (November 1941): 239
58 *Ibid.*, p. 238
59 *Canadian Unionist* (September 1941), p. 87
60 Memo on the conduct of the vote, file 2900, J.L. Cohen papers, PAC
61 *Ibid.*
62 E. Long, 10 November 1941, letter to J.R. Mutchmor, J.R. Mutchmor personal correspondence 1941, United Church Archives
63 Account of the vote, file 2900, J.L. Cohen papers, PAC
64 *Financial Post*, 15 November 1941
65 R. Livett, 7 November 1941, letter to N. McLarty, file 2900, J.L. Cohen papers, PAC
66 Silby Barrett, 7 November 1941, letter to N. McLarty, *ibid.*

67 *Financial Post*, 15 November 1941
68 *Debates*, House of Commons, 7 November 1941, p. 4167
69 J.R. Mutchmor, 7 November 1941, letter to N. McLarty, J.R. Mutchmor personal correspondence 1941, United Church Archives
70 'Further Information to Shareholders on the Kirkland Lake Situation,' 15 November 1941, J.R. Mutchmor personal correspondence 1941, United Church Archives
71 *Financial Post*, 15 November 1941
72 Memo on the conduct of the vote, file 2900, J.L. Cohen papers, PAC
73 Long, 19 November 1941, letter to J.R. Mutchmor, J.R. Mutchmor personal correspondence 1941, United Church Archives
74 *Debates*, House of Commons, 10 November 1941, p. 4217
75 *Ibid.*, p. 4218
76 *The Globe and Mail*, 15 November 1941
77 M.W. Summerhayes and K.C. Gray, 10 November 1941, letter to WLMK, vol. 152, PMO files, King papers, PAC
78 *Northern Miner*, 15 November 1941
79 *Financial Post*, 15 November 1941; *The Globe and Mail*, 19 November 1941
80 *Labour Gazette* (Ottawa 1941), p. 1492. The breakdown of the vote was as follows:

Name of mine	Number of eligible voters	Number for strike	Number voting No	Spoiled
Bidgood	128	69	48	8
Brock Gold Mines*	31	12	2	2
Golden Gate*	55	10	39	2
Kirkland Lake Gold	326	232	72	3
Lake Shore	1037	680	273	17
Macassa	292	180	95	3
Morris-Kirkland*	42	18	14	–
Sylvanite	400	233	126	14
Teck-Hughes	550	413	109	1
Toburn	204	105	83	5
Upper Canada*	195	75	95	9
Wright Hargreaves	1073	698	298	14
Total	4333	2725	1254	78

*Four companies with clear majority voting against strike
81 *Toronto Star*, 11 November 1941
82 Grube, 'The Strike at Kirkland Lake,' p. 300
83 Tom McGuire, 1 November 1941, letter to J.L. Cohen, file 2894, J.L. Cohen papers, PAC

84 *Northern Miner*, 13 November 1941
85 Carlin, transcript, p. 45
86 *Ibid.*, p. 20

CHAPTER 6

1 No relation to Tom McGuire, administrator of the union
2 *Northern News*, 4 November 1941
3 I. Avakumovic, *The Communist Party of Canada* (Toronto: McClelland and Stewart Ltd 1975), p. 144. Dorise Nielson was elected as a 'Unity' MP but was also a prominent member of the LPP (i.e. the Communist Party).
4 *Northern News*, 4 November 1941
5 'The Labour Situation in Kirkland Lake: An Important Statement,' vol. 196, Canadian Labour Congress (CLC) papers, Public Archives of Canada (PAC)
6 *Hamilton Spectator*, 11 November 1941; *Northern Miner*, 13 November 1941
7 *The Globe and Mail*, 19 November 1941
8 *Canadian Unionist* (October 1940), p. 125; (February 1941), p. 214
9 See chapter 2 regarding the Peck Rolling Mills dispute
10 *Canadian Unionist* (March 1941), p. 240
11 *Ibid.* (July 1941), p. 1
12 *Ibid.* (August 1941), p. 53
13 *Ibid.* (September 1941), p. 76
14 A.R. Mosher, 10 October 1941, letter to Tom McGuire, vol. 99, CLC papers, PAC
15 *Canadian Unionist* (October 1941), p. 103
16 Stenographer's minutes, CCL executive meeting, 21 October 1941, vol. 99, CLC papers, PAC
17 *Ibid.*
18 *Ibid.*
19 *Ibid.*
20 Stenographer's minutes, CCL executive council meeting, 22 October 1941, vol. 99, CLC papers, PAC
21 Paul MacEwan, *Miners and Steelworkers* (Toronto: Samuel Stevens, Hakkert and Co. 1976), chapter 17. The Nova Scotia miners' illegal slowdown occurred when the union executive agreed to a contract without holding a ratification vote of the miners. The unrest also reflected the miners' inadequate wages and working conditions since the 1930s.
22 'The Issue at Kirkland Lake,' *CCL Bulletin*, no. 13 (15 November 1941), vol. 196, CLC papers, PAC
23 *CCL Bulletin*, no. 15 (22 November 1941), *ibid.*
24 *Canadian Unionist* (October 1941), p. 107

25 Grant MacNeil Correspondence, file 2, vol. 102, national CCF papers, PAC; series III, minutes of the Provincial Executive and Council 1934–52, Ontario CCF papers, Queen's University

26 *Debates*, House of Commons, 18 November 1940, p. 183

27 Grant MacNeil, 15 March 1941, letter to Mr Dalton, vol. 102, national CCF papers, PAC

28 MacNeil, 14 March 1941, letter to national office, *ibid.*

29 *Debates*, House of Commons, 19 November 1940, pp. 216–17

30 *Ibid.*, 4 November 1941, p. 4072

31 *Ibid.*, p. 4074

32 *Ibid.*, 14 November 1941, p. 4410

33 Bob Carlin, 10 November 1941, letter to J.P. Harris et al., file 2900, J.L. Cohen Papers, PAC

34 William Simpson, 10 November 1941, letter to Norman McLarty, *ibid.*

35 Norman McLarty, 12 November 1941, letter to Bob Carlin, *ibid.*

36 Carlin, 12 November 1941, letter to Norman McLarty, *ibid.*; E. Long, 13 November 1941, letter to J.R. Mutchmor, J.R. Mutchmor personal correspondence 1941, United Church Archives

37 The union's delegation was comprised of Bob Carlin, Larry Sefton, Jock Brodie, and William Simpson (*Montreal Gazette*, 18 November 1941).

38 J.L. Cohen, 15 November 1941, letter to Norman McLarty, file 2900, J.L. Cohen papers, PAC

39 *Montreal Gazette*, 18 November 1941

40 *Toronto Star*, 13 November 1941

41 George Grube, 'The Strike at Kirkland Lake,' *Canadian Forum* (January 1942): 301

42 Carlin, financial secretary, Local 240, transcript of taped interview with the author at Gowganda, August 1973, p. 32

43 Grube, 'The Strike at Kirkland Lake,' p. 301

44 From the pamphlet, 'Further Information to Shareholders on the Kirkland Lake Labour Situation,' J.R. Mutchmor general correspondence 1941, United Church Archives

45 McLarty, 3 December 1941, letter to William Lyon Mackenzie King (WLMK), primary correspondence, vol. 310, King papers, PAC

46 *Ibid.*

47 Carlin, transcript, p. 32

48 McLarty, 3 December 1941, letter to WLMK, primary correspondence, vol. 310, King papers, PAC. This version also appeared in *The Globe and Mail*, 19 November 1941

49 *Montreal Star*, 20 November 1941

50 *Toronto Star*, 19 November 1941

CHAPTER 7

1 *Toronto Star*, 19 November 1941
2 Kay Carlin, transcript of taped interview with the author at Gowganda, August 1973, p. 2
3 *Toronto Star*, 19 November 1941
4 *Ibid.*, 21 November 1941
5 *Timmins Press*, 22 November 1941
6 Kay Carlin, transcript, p. 4
7 Bob Carlin, financial secretary, Local 240, transcript of taped interview with the author at Gowganda, August 1973, p. 29
8 *Ibid.*, p. 56. 'And as to women, my experience ... has been that if you've got backward women you never have a strong union. It's like trying to, well, go to bat with one arm tied behind your back. The woman has got a vote, she's got a voice, and well ... if she ... is steeped in ignorance regarding the operation and function of the union, how the hell can you expect anything different in a time of crisis?'
9 Leaflet, 'From One Woman to Another,' file 2980, J.L. Cohen papers, Public Archives of Canada (PAC)
10 *Ibid.*
11 Kay Carlin, transcript, p. 2
12 *Ibid.*, pp. 3–4. 'The weather conditions severe, the temperature according to official records, dropping to a low of 48 degrees below zero at Kirkland Lake and 52 degrees below zero at King Kirkland last night,' Commissioner Stringer, 10 January 1942 to Attorney General Conant, file 11.8 Provincial Police (Strikes 1942), Ontario Attorney General's papers, Archives of Ontario.
13 Kay Carlin, transcript, pp. 4, 6, 7
14 *Ibid.*, p. 6
15 *Northern News*, 19 August 1941
16 J.R. Mutchmor, 13 December 1941, letter to Rev. S.J. Bridgette, J.R. Mutchmor personal correspondence, 1941, United Church Archives
17 P. Bennett, 10 November 1941, letter to Mitchell Hepburn, public correspondence, 1941, Hepburn papers, Public Archives of Ontario (PAO)
18 E. Long, 30 October 1941, letter to J.R. Mutchmor, J.R. Mutchmor personal correspondence, 1941, United Church Archives
19 Bennett, 10 November 1941, letter to Mitchell Hepburn, public correspondence, 1941, Hepburn papers, PAO
20 *United Church Observer*, 15 February 1942, E. Long, 'The Church in Striketown'
21 Long, 30 October 1941, letter to J.R. Mutchmor, J.R. Mutchmor personal correspondence, 1941, United Church Archives

22 'Strike Over – Assistance Still Needed!,' *CCL Bulletin*, no. 6, 18 February
   1942, vol. 196, Canadian Labour Congress (CLC) papers, PAC
23 'As the Collector Sees It,' unknown author, a collector and credit manager in
   Kirkland Lake, file 2780, J.L. Cohen papers, PAC
24 Untitled speech, file 2900, *ibid.*
25 The middle class refers to half of the business community, professionals, the
   mine management personnel, most clergymen, and most offices workers in
   the mines who were not unionized.
26 *The Globe and Mail*, 29 November 1941
27 *Ibid.*, 2 December 1941
28 *Ibid.*, 3 December 1941
29 Rex Lucas, *Minetown, Milltown and Railtown* (Toronto: University of Toronto
   Press 1971), pp. 316, 319
30 *Ibid.*, p. 319
31 Long, 16 December 1941, letter to J.R. Mutchmor, J.R. Mutchmor personal
   correspondence, 1941, United Church Archives
32 Bob Carlin, transcript, p. 68
33 *Ibid.*, p. 65; Wayne Roberts, ed., *Miner's Life* (Hamilton: McMaster University 1979), p. 6
34 Handwritten summary of the union's activities, 9 September 1941, Larry Sefton papers. The union apparently met with local church leaders to try to
   influence them. The above-mentioned document concludes: 'This summary of
   the union's activities was presented September 9th, to a meeting in United
   Church of the ministers of the Presbyterian and United Church from Haileybury to Timmins. Presented along with the necessary verbal additions to justify our past and present actions with the hope in mind that the clergy would
   support the fundamental rights of so many of their people which the union is
   trying to establish.'
35 Long, 13 November 1941, letter to J.R. Mutchmor, J.R. Mutchmor personal
   correspondence, 1941, United Church Archives
36 J. Boyd, 16 December 1941, letter to J.R. Mutchmor, J.R. Mutchmor personal correspondence, 1941, United Church Archives
37 'Report of the Temiskaming Presbytery re: Kirkland Lake Labour Situation,'
   25 September 1941, Board of Evangelism and Social Service, box 20, general
   files, United Church Archives
38 Long, 10 November 1941, letter to J.R. Mutchmor, J.R. Mutchmor personal
   correspondence, 1941, United Church Archives
39 Mutchmor, 19 December 1941, letter to R.J. Scott, *ibid.*
40 Mutchmor, 19 December 1941, letter to E. Long and J. Boyd, *ibid.*
41 J. Coburn, 12 December 1941, letter to W. Cullis, *ibid.*

42 *Ibid.*
43 Resolution of the Toronto West Presbytery, *ibid.*
44 Coburn, 24 December 1941, letter to W. Cullis, *ibid.*
45 Mutchmor, 19 December 1941, letter to E. Long and J. Boyd, *ibid.*
46 Boyd, 16 December 1941, letter to J.R. Mutchmor, *ibid.*
47 Arnold Brown, 'A Newspaper Fights a Miners' Strike,' *Canadian Forum* (January 1942): 301
48 Richard Pearce, 16 September 1941, letter to Mitchell Hepburn, Hepburn papers, 1941, Archives of Ontario (PAO)
49 *Northern News*, 6 January 1942
50 *Ibid.*, 5 September 1941
51 For information about Robinson's political background see Vernon Jensen, *Nonferrous Metals Industry Unionism* (New York: Cornell University 1954). Richard Pearce (editor, *Northern Miner*) 18 June 1942 letter to Attorney General Conant, file 11.8, Provincial Police (Strikes 1942), Ontario Attorney General's papers, Archives of Ontario. He discusses the libel suit saying, 'We had to go to considerable pains to protect ourselves ... You will be interested in knowing that in examination for discovery Robinson admitted all the connections I brought to your attention, and some others. By arrangement the action was dismissed.'
52 Bob Carlin, transcript, p. 60
53 'Kirkland Lake Bulletin,' 5 December 1941, vol. 196, CLC papers, PAC
54 Kay Carlin, as recounted to Elaine Sefton
55 Mrs L.E. Hansmen, 28 November 1941, letter to Mitchell Hepburn, public correspondence, 1941, Hepburn papers, PAO
56 D.C. Masters, *The Winnipeg General Strike* (Toronto: University of Toronto Press 1973), p. 51
57 Radio speech, Alex Harris, 18 December 1941, J.N. 'Pat' Kelly papers, McMaster University Archives
58 Leaflet, 'Company Money Back of Workers' Council,' file 2900, J.L. Cohen papers, PAC
59 'Doc' Ames was approached by the management of one of the mining companies to lead its employee committee. He refused and supported the miners in the strike. Taped conversation with the author, August 1973
60 Officer McDougall's 13th Report, 31 October 1941, file 11.8 Provincial Police (Strikes 1942), Ontario Attorney General's papers, Archives of Ontario: 'In the event of a strike the mines concerned will keep open and any man that is willing to work may do so.'
61 *The Globe and Mail*, 15 November 1941
62 Long, 30 October 1941, letter to J.R. Mutchmor, J.R. Mutchmor personal correspondence, 1941, United Church Archives

63 *Ibid.*, 10 November 1941
64 *Northern Miner*, 13 November 1941, p. 137
65 Tom McGuire, 20 November 1941, letter to J.L. Cohen, file 2900, J.L. Cohen papers, PAC
66 *Toronto Star*, 19 November 1941
67 'Kirkland Lake Bulletin,' no. 1, 16 December 1941, vol. 196, CLC Papers, PAC
68 *Ibid.*, 2 January 1942
69 *The Globe and Mail*, 28 November 1941
70 *Ibid.*, 2 December 1941
71 Radio speech, Alex Harris, 18 December 1941, Kelly papers, McMaster University Archives
72 Report, file 263, vol. 414, Canada, Department of Labour papers, PAC
73 Bob Carlin, transcript, p. 63
74 *Labour Gazette* (1941), p. 155; *Labour Gazette* (1942), p. 278
75 Staff Inspector Doyle, 20 January 1942, letter to Commissioner Stringer, file 11.8 Provincial Police (Strikes 1942), Ontario Attorney General's papers, Archives of Ontario. The breakdown of numbers of people working was recorded for that date as follows:

Mine	Miners	Staff	Total
Bidgood	94	16	110
Macassa	109	24	133
Kirkland Lake Gold	89	14	103
Teck-Hughes	136	21	157
Lake Shore	301	49	350
Wright-Hargreaves	432	76	508
Sylvanite	172	27	199
Toburn	109	15	124
	1442	242	1684

76 *Debates*, House of Commons, 18 November 1940 (Clarie Gillis), p. 182
77 M. Zwelling, *The Strikebreakers* (Toronto: New Press 1972), p. 8
78 *Ibid.*, p. 7
79 Stuart Jamieson, *Times of Trouble: Labour Unrest and Industrial Conflict in Canada 1900–66*, Study No. 22 (Ottawa: Task Force on Labour Relations 1968), p. 53
80 V.H. Emery (manager, Kirkland Lake Gold Mine), 12 November 1941 letter to Conant, file 11.8 Provincial Police (Strikes 1942), Ontario Attorney General's papers, Archives of Ontario: 'We want to give any men who wish to work the opportunity to do so. However we do not want to see our loyal men or their families subjected to abuse. We are anxious for adequate protection to see law and order is maintained in the camp.'

81 Desmond Morton, 'Aid to Civil Power: The Canadian Militia in Support of Social Order,' *Canadian Historical Review* (December 1970): 407–25

82 *Northern Miner*, 13 November 1941, p. 137

83 Long, 30 October 1941, letter to J.R. Mutchmor, J.R. Mutchmor personal correspondence, 1941, United Church Archives

84 Fear that Hepburn might lash out at the federal government and adversely affect the outcome of the Welland and South York by-elections contributed to the federal government's silence at the end of the strike. It refused to intervene to prevent company discrimination as the workers returned to work. See chapter 10 and also Reid Robinson and W. Simpson, 5 February 1942, telegram to WLMK prime minister's comments on telegram, prime minister's office (PMO) file, vol. 152, King papers, PAC

85 *The Globe and Mail*, 2 December 1941

86 E.T. Doyle (staff inspector) OPP, 18 November 1941, letter to Commissioner Stringer, file 11.8 Provincial Police (Strikes 1942), Ontario Attorney General's papers, Archives of Ontario. He said that Healey, the manager of Wright-Hargreaves had 'reliable' information that the power lines might be threatened.'

87 *Toronto Star*, 21 November 1941

88 *Timmins Press*, 22 November 1941

89 *Ibid.*

90 *Ibid.*

91 *The Globe and Mail*, 24 November 1941

92 *Ibid.*

93 Annual Report of the OPP (1941), *Ontario Sessional Papers* (1941), p. 34

94 M.W. Summerhayes, 22 November 1941, letter to Mitchell Hepburn, public correspondence, 1941, Hepburn papers, PAO

95 *The Globe and Mail*, 24 November 1941; Annual Report of the OPP (1941), p. 34

96 *The Globe and Mail*, 24 November 1941; *Ottawa Morning Journal*, 25 November 1941

97 *The Globe and Mail*, 25 November 1941

98 Annual Report of the OPP (1941), p. 35

99 M.W. McBain, clerk, Teck Township Council, 25 November 1941, letter to Mitchell Hepburn, Public Correspondence, 1941, Hepburn Papers, PAO

100 *The Globe and Mail*, 25 November 1941

101 *Ibid.*, 24 November 1941

102 D.G.H. Wright, 24 November 1941, letter to Mitchell Hepburn, public correspondence, 1941, Hepburn papers, PAO

103 S. Lapede, 27 November 1941, letter to Mitchell Hepburn, *ibid.*

104 W.G. Nixon, MPP, 25 November 1941, letter to Mitchell Hepburn, *ibid.*

105 B.A. Haller, 24 November 1941, letter to Mitchell Hepburn, *ibid.*
106 E. Glanfield, 24 November 1941, letter to Mitchell Hepburn, *ibid.*
107 *The Globe and Mail*, 26 November 1941, 27 November 1941
108 Case of Cecil Peters, 6 December 1941, file 2903, J.L. Cohen papers, PAC
109 Annual Report of the OPP (1941), p. 34
110 *The Globe and Mail*, 5 December 1941
111 Kay Carlin, transcript, p. 4
112 Annual Report of the OPP (1941), p. 34
113 See note 86
114 *The Globe and Mail*, 5 December 1941; 16 December 1941
115 Case of John Broughton, file 2903, J.L. Cohen papers, PAC
116 Bob Carlin, transcript, p. 35
117 *Ibid.*, p. 36
118 Jamieson, *Times of Trouble*, p. 284
119 Conant, 25 February 1941, letter to M.W. Hotchkin (Kirkland Lake Mine Operators' Committee), file 11.8, Provincial Police, (Strikes 1942), Ontario Attorney General's papers, Archives of Ontario.
120 Kay Carlin, transcript, p. 3
121 J. Coburn, 19 December 1941, letter to E. Long and J. Boyd, J.R. Mutchmor personal correspondence, 1941, United Church Archives
122 Annual Report of the OPP (1941), p. 34
123 M.W. Hotchkin, 18 February 1942, letter to A. Conant, private correspondence, 1942, Hepburn Papers, PAO. (My emphasis.)
124 Hotchkin, 18 February 1942, letter to Mitchell Hepburn, *ibid.*
125 Herbert Quinn, *The Union Nationale* (Toronto: University of Toronto Press 1963), p. 94
126 J.L. Cohen, 8 December 1941, letter to Reid Robinson, file 2900, J.L. Cohen papers, PAC
127 William Simpson, 29 November 1941, letter to William Lyon Mackenzie King (WLMK), vol. 318, WLMK primary correspondence, King papers, PAC
128 McGuire, 7 December 1941, letter to Cohen, file 2900, J.L. Cohen papers, PAC
129 Cohen, 8 December 1941, letter to Reid Robinson, *ibid.*
130 *CIO News*, 8 December 1941
131 *Steel Labor*, 19 December 1941
132 H.A. Logan, *Trade Unions in Canada* (Toronto: Macmillan Co. of Canada Ltd 1948), p. 546; *CCL Convention Proceedings 1942*, pp. 18–19, vol. 212, CLC Papers, PAC
133 Cohen, 8 December 1941, letter to Reid Robinson, file 2900, J.L. Cohen papers, PAC

134 Private and public correspondence, 1941–42, Hepburn papers, PAO; PMO file, vol. 152, King papers, PAC

135 *The Globe and Mail*, 1 December 1941

136 *CCL Bulletin* no. 1, 16 December 1941, vol. 196, CLC papers, PAC

137 Cohen, 8 December 1941, letter to Reid Robinson, file 2900, J.L. Cohen papers, PAC

138 *The Globe and Mail*, 5 December 1941, 6 December 1941

139 *Ibid.*, 24 November 1941

140 Coburn, 24 December 1941, letter to W. Cullis, J.R. Mutchmor personal correspondence, 1941, United Church Archives

141 *Annual Report 1942* (Board of Evangelism and Social Service), p. 102, box 24, United Church Archives

142 *Annual Report 1941* (Board of Evangelism and Social Service), p. 55, *ibid.*

143 *Annual Report 1942* (Board of Evangelism and Social Service), p. 101

144 *Ibid.*

145 *Toronto Star*, 10 December 1941

146 WLMK typescript diary, 23 October 1941, vol. 89, King papers, PAC

147 Coburn, 3 January 1942, letter to J. Atkinson, J.R. Mutchmor personal correspondence, 1941, United Church Archives

148 *Ibid.*

149 Coburn, 19 December 1941, letter to E. Long and J. Boyd, J.R. Mutchmor personal correspondence, 1941, United Church Archives

150 *Ibid.*

151 Michael Bliss, 'The Methodist Church and World War I,' *Canadian Historical Review* XLIX (September 1968): 3

152 J.R. Mutchmor, 19 December 1941, letter to E. Long and J. Boyd, J.R. Mutchmor personal correspondence, 1941, United Church Archives. Both the church leaders and the mine operators' views of the war were similar, while their views on the cause of labour unrest were diametrically opposed. King refused to resolve the labour problem in the way suggested by church and labour leaders. By his refusal he alienated organized labour, which as an interest group opposed Meighen's conscription policy. Instead, he assisted a pro-conscription group of mining 'patriots' to win the Kirkland Lake strike over their employees.

153 Coburn, 19 December 1941, letter to E. Long and J. Boyd, *ibid.* King was worried by the Conservative call for a National Government.

154 *The Globe and Mail*, 6 December 1941

155 Norman McLarty, 3 December 1941, letter to WLMK, vol. 310, WLMK primary correspondence, King papers, PAC

156 A.A. Macleod, 19 November 1941, letter to J.L. Cohen, file 2900, J.L. Cohen papers, PAC

157 B. Leavens, 21 March 1941, letter to Grant MacNeil, vol. 102, national Co-operative Commonwealth Federation (CCF) papers, PAC
158 *Debates*, House of Commons, 18 November 1940 (Clarie Gillis), p. 179
159 J. Tough (Windsor Central CCF Club), 14 December 1941, letter to David Lewis, vol. 188, national CCF papers, PAC
160 David Lewis, 27 December 1941, letter to Bob Carlin, *ibid.*

CHAPTER 8

1 Neil McKenty, *Mitch Hepburn* (Toronto: McClelland and Stewart 1967), p. 235
2 *Ibid.*, p. 236
3 G. Grube, 'The Strike at Kirkland Lake,' *Canadian Forum* (January 1942): 299
4 McKenty, *Mitch Hepburn*, p. 235
5 Stuart Jamieson, *Industrial Relations in Canada* (Toronto: Macmillan and Co. Ltd 1973), p. 116
6 'Promoting Patriotism,' *Canadian Forum* (February 1942): 326. This group, which included the mining interests, supported Meighen for the leadership of the Conservative Party and conscription. *Canadian Forum* put forward the view that 'Mr. King, largely out of fear of this vocal minority has already gone a long way to sacrifice national unity by his disastrous labour policy.' This group, centred around the *Globe and Mail*, was the focus 'of both the pro-conscription and anti-labour drives.' Mr Wright who owned the *Globe and Mail* also owned a large share of Wright-Hargreaves mine at Kirkland Lake. The Kirkland Lake dispute was 'the outstanding example of the weakness of the government's labour policy.' Interestingly, while King was trying to conciliate this powerful group of mining 'patriots,' they were capitalizing on the growing public discontent with the King government's conduct of the war.
7 William Lyon Mackenzie King (WLMK) typescript diary, 5 December 1941, vol. 89, King papers, PAC
8 *Ottawa Morning Journal*, 7 November 1941
9 *Toronto Star*, 7 November 1941
10 *Ottawa Morning Journal*, 7 November 1941
11 For a discussion of the philosophical underpinnings of WLMK's consistent philosophy of labour relations see Paul Craven, *'An Impartial Umpire': Industrial Relations and the Canadian State 1900–11* (Toronto: University of Toronto Press 1980); Reginald Whitaker, 'The Liberal Corporatist Ideas of Mackenzie King,' *Labour/Le Travailleur* 2 (1977)

12 J.R. Mutchmor, 19 December 1941, letter to J. Boyd and E. Long, J.R. Mutchmor personal correspondence, 1941, United Church Archives

13 Angus MacInnis and M.J. Coldwell, 27 November 1941, letter to WLMK, vol. 188, national Co-operative Commonwealth Federation (CCF) papers, Public Archives of Canada (PAC)

14 David Lewis, 27 and 28 July and 3 August, 1941, letters to Tom McGuire, *ibid.* Lewis aided Local 240 in obtaining a passport for its recording secretary, Larry Sefton. He obtained American money (the release of which required the permission of the Foreign Exchange Control Board) for the union's delegates to Mine Mill's international convention in Joplin, Missouri. Lewis wrote: 'I was very glad to do this little chore for you and if there is anything further we can do, please do not hesitate to write me.' Lewis' activities are now verified in David Lewis, *The Good Fight: Political Memoirs 1909–1958* (Toronto: Macmillan 1981).

15 Norman McLarty, 4 December 1941, letter to M.J. Coldwell, WLMK primary correspondence, vol. 310, King papers, PAC

16 *Ibid.*

17 Donald J.M. Brown, *Interest Arbitration*, Study no. 18 (Ottawa: Task Force on Labour Relations 1968), p. 252

18 WLMK, 16 December 1941, letter to M.J. Coldwell, vol. 188, national CCF papers, PAC

19 J.R. Mutchmor, 9 December 1941, letter to E. Long, J.R. Mutchmor Personal Correspondence, 1941, United Church Archives. 'If they lost because of Federal government inaction, labour both in the CIO and AFL groups, as well as organized workers will know that they cannot put any confidence in Mr. Mackenzie King and his Cabinet. The great army of men in overalls will be quite convinced that the right of collective bargaining has ceased to exist in Canada.'

20 WLMK, 16 December 1941, letter to M.J. Coldwell, vol. 188, national CCF papers, PAC

21 W.J. Turnbull, 1 December 1941, memo to WLMK, WLMK memoranda and notes 1940–50, vol. 299, King papers, PAC

22 WLMK typescript diary, 2 December 1941, 5 December 1941, vol. 89, King papers, PAC

23 WLMK, 6 December 1941, letter to William Simpson, vol. 318, WLMK primary correspondence, King papers, PAC

24 E. Long, 10 December 1941, letter to J.R. Mutchmor, J.R. Mutchmor Personal Correspondence, 1941, United Church Archives

25 Norman McLarty, 3 December 1941, letter to WLMK, vol. 310, WLMK primary correspondence, King papers, PAC

26 *Ibid.*

27 I.M. Christie, *The Liability of Strikers in the Law of Tort* (Kingston: Queen's University 1967). I am indebted to Richard MacDowell for clarification on this point.

28 'Ontario: Proceedings of the Select Committee re Bargaining Between Employers and Employees' (Toronto 1943), Ontario Legislative Library

29 Richard Polenberg, *War and Society* (New York: J.B. Lippincott Co. 1972), pp. 7, 20, 103

30 Norman McLarty, 3 December 1941, letter to WLMK, vol. 310, WLMK primary correspondence, King papers, PAC

31 *Ibid.*

32 *Ibid.*

33 Bulletin no. 1, 16 December 1941, Kirkland Lake Strike Committee, vol. 196, Canadian Labour Congress (CLC) papers, PAC

34 Bulletin no. 2, 27 December 1941, *ibid.*

35 Bulletin no. 3, 8 January 1942, *ibid.*

36 *Northern News*, 16 January 1942; *The Globe and Mail*, 14 January 1942

37 Bulletin no. 1, 16 December 1941, Kirkland Lake Strike Committee, vol. 196, CLC Papers, PAC

38 Daniel Coates, 'Organized Labor and Politics in Canada: The Development of a National Labor Code' (PHD dissertation, Cornell University 1973), p. 78

39 *Ibid.*, p. 101

40 Charlie Millard, 26 November 1941, letter to WLMK, vol. 312, WLMK primary correspondence, King papers, PAC

41 A.R. Mosher, 16 December 1941 press release file, press releases 1940–41, box 11 (unprocessed material), CLC papers, PAC. The press release was made in response to press requests for comment on the appointment of Mitchell. 'I have no hesitation in assuring Mr Mitchell that he will have the full co-operation of the CCL in every effort which may be made to establish a consistent and progressive labour policy for the Dominion, and wish him every success in carrying on the duties of the office he has assumed. Organized labour will judge Mr Mitchell by his performance as Minister of Labour however rather than by his past activities. No one knows better than he does the importance of establishing and retaining the confidence of organized labour in the department over which he presides, and in the government as a whole. The workers seek no favours; they do, however, insist upon a square deal, and can be expected to give of their best only under conditions which ensure just treatment by both government and employers.'

42 WLMK typescript diary, 14 December 1941, 16 December 1941, vol. 89, King papers, PAC

43 WLMK, 17 December 1941, letter to Walter Willoughby, vol. 152, prime minister's office (PMO) files, King papers, PAC

44 *The Globe and Mail*, 24 November 1941
45 Humphrey Mitchell, 9 January 1942, letter to William Simpson, File 2900, J.L. Cohen papers, PAC
46 William Simpson, n.d., letter to Humphrey Mitchell, *ibid.*
47 *Toronto Star*, 14 January 1942
48 J.L. Cohen, 21 January 1942, letter to Reid Robinson, file 2900, J.L. Cohen papers, PAC
49 Humphrey Mitchell, 21 January 1942, letter to William Simpson, *ibid.*
50 A.R. Mosher, 24 January 1942, letter to Humphrey Mitchell, *ibid.*
51 Cohen, 26 January 1942, letter to A.D. McNeil (assistant to Reid Robinson), *ibid.*
52 Humphrey Mitchell, 21 January 1942, letter to William Simpson, *ibid.*; *Debates*, House of Commons, 26 January 1942, p. 19. Clarie Gillis enquired if the mine operators had responded to the arbitration proposal. McLarty, in Mitchell's absence, replied that the minister had received several communications of acceptance from the union, though now *there were certain qualifications of procedure*. The mine owners had made no *formal* reply to the government.
53 *Northern News*, 2 January 1942; William Simpson, 14 January 1942, letter to WLMK, vol. 152, PMO files, King papers, PAC
54 *The Globe and Mail*, 13 January 1942
55 W.J. Turnbull, 28 January 1942, note to WLMK, vol. 152, PMO files, King papers, PAC
56 WLMK typescript diary, 4 February 1942, vol. 90, King papers, PAC
57 *Ibid.*
58 Bulletin no. 5, 4 February 1942, Kirkland Lake Strike Committee, vol. 196, CLC papers, PAC
59 *Toronto Star*, 6 February 1942
60 *Winnipeg Manitoba Commonwealth*, 20 March 1942
61 G. Grube, 'Defeat in Kirkland Lake,' *Canadian Forum* (March 1942): 359
62 W.P. St Charles, 18 February 1942, letter to WLMK, vol. 152, PMO files, King papers, PAC
63 Grube, 'The Strike at Kirkland Lake,' p. 299. Reg Whitaker, 'The Liberal Corporatist Ideas of Mackenzie King,' *Labour/Le Travailleur* 2 (1977). His analysis of King's ideas about labour relations is in accord with my analysis of King's *administration* of labour policy during the war.

CHAPTER 9

1 Bob Carlin, financial secretary, Local 240, transcript of taped interview with the author at Gowganda, August 1973, p. 43

2 Memo to the prime minister, 11 November 1941, file 2817, vol. 276, William Lyon Mackenzie King (WLMK) memoranda and notes, King papers, Public Archives of Canada (PAC)
3 *Winnipeg Free Press*, 22 January 1942
4 Carlin, transcript, p. 43
5 *Toronto Star*, 6 February 1942
6 *Toronto Star*, 10 February 1942; handwritten minutes, 27 January 1942, National Strike Committee, file – Kirkland Lake Mine and Mill Workers' Union National Strike Committee, box 7, unprocessed material, CLC papers, PAC. Some CCL money was spent to try to defeat Mitchell in the Welland by-election.
7 Note on telegram from Reid Robinson to William Simpson, 5 February 1942 to WLMK, vol. 152, prime minister's office (PMO file), King papers, PAC
8 *Toronto Star*, 6 February 1942
9 N. McLarty, 3 December 1941, letter to WLMK, vol. 310, primary correspondence, King papers, PAC
10 Expense sheet 9–14 February 1942, file 2900, J.L. Cohen papers, PAC
11 WLMK, 6 February 1942 letters to Local 4481 UMWA, and locals 2251 and 1231 USWA, vol. 152, PMO file, King papers, PAC
12 Bryce Stewart, 2 February 1942, memo to W.J. Turnbull (principal secretary to the prime minister), *ibid.* See also Paul MacEwan, *Miners and Steelworkers* (Toronto: Hakkert and Co. 1976), chapter 17. The Nova Scotia miners went on an illegal strike in 1941.
13 Company reports, file 263, vol. 414, Canada Department of Labour
14 *Canadian Tribune*, 9 January 1943
15 'Remember Kirkland Lake,' leaflet, 16 February 1942, vol. 196, Canadian Labour Congress (CLC) papers, PAC
16 Carlin, personal interview with the author at Gowganda, August 1973
17 *London Free Press*, 18 February 1942
18 *Debates*, House of Commons (McLarty) 13 February 1942, p. 569
19 *Ibid.* (Mitchell), p. 922
20 *Ibid.* (Nielson)
21 *Ibid.* (Mitchell), p. 780
22 Commissioner Stringer, 11 February 1942, letter to Conant, file 11.8 Provincial Police (Strikes 1942), Ontario Attorney General's papers. Although federal Deputy Labour Minister Bryce Stewart had suggested that the striking miners be taken back on a seniority basis, 'the department did not propose to participate any further in the matter.' Stringer then related the policy of the mine operators with regard to rehiring. 'Mr Hotchkin (manager of Toburn mine) also pointed out that the mine operators do not propose to let down

loyal and trusted employees who stayed with them throughout the crisis to make room for striking miners and would take men back only if and when there were positions open for them. He said there would be no statement issued by the mine operators dealing with the present situation.'

23 Executive Council Meeting, CCL, 26 February 1942, vol. 99, CLC papers, PAC
24 *Debates*, House of Commons (Gillis), 18 March 1942, p. 1415
25 *Ibid.* (Hanson), 2 March 1942, p. 926
26 Local 240, 9 February 1944, letter to all MPs, vol. 28, CLC papers, PAC
27 John Lang, 'A Lion in a Den of Daniels: A History of the IUMMSW in Sudbury, Ontario 1942–62' (MA dissertation, University of Guelph 1973), p. 46
28 *Debates*, House of Commons (MacInnis), 5 February 1942, p. 339

CHAPTER 10

1 *Northern Miner*, 11 September 1941; *The Globe and Mail*, 19 November 1941
2 It was reported that the gold index was down to a new low in 1941. It resulted from speculation about a change in government policy that would give base metals a priority over gold mining in the allotment of steel requirements. (*The Globe and Mail*, 19 December 1941).
3 *Ibid.*, 8 January 1942
4 *Northern Miner*, 11 September 1941
5 *Ibid.*, 13 November 1941
6 *Vancouver Province*, 18 February 1942
7 *Northern Miner*, 19 March 1942
8 *Ibid.*, 13 November 1942
9 *Ottawa Morning Journal*, 3 December 1941
10 *Toronto Financial Post*, 4 March 1941
11 *Timmins Press*, 21 November 1941
12 CCL convention proceedings, 1942 stenographic report, p. 368, vol. 212, Canadian Labour Congress (CLC) papers, Public Archives of Canada (PAC)
13 *Northern Miner*, 19 March 1942
14 *Sudbury Star*, 3 March 1942
15 *Ibid.*
16 Alex Harris (Kirkland Lake Gold Mine Operators' Committee), 17 July 1946, letter to H.G. Hilton (Stelco), J.N. 'Pat' Kelly papers, McMaster University Archives
17 *Hamilton Spectator*, 8 December 1943
18 CCL executive meeting, 26 February 1942, vol. 99, CLC papers, PAC. There was some criticism that Millard took too firm a stand and some suggestion that had the union worked through the employees' committees, it might have

gained *de facto* recognition. H.A. Logan, *Trade Unions in Canada* (Toronto: The Macmillan Co. of Canada Ltd 1948), p. 546. In view of the past history of labour relations in the mining industry, and the distrustful state of labour/government relations, this strategy was never given any serious consideration.

19 *Ibid.* The most critical unions were influenced by communists within their ranks who distrusted the CCL leadership for political reasons, because it was oriented to the CCF.

20 *Hamilton Spectator*, 8 December 1943

21 *Oshawa Times Gazette*, 11 December 1943

22 CCL executive meeting, 26 February 1942, vol. 99, CLC papers, PAC

23 Kay Carlin, transcript of taped interview with the author at Gowganda, August 1973, p. 8

24 *Debates*, House of Commons, 25 February 1942, p. 820

25 Local 240 executive, 9 February 1944, letter and resolution to all MPs and MPPs, vol. 28, CLC papers, PAC

26 B.T. Doherty, 6 April 1941, letter to A. McNamara, *ibid.*

27 Vernon Jensen, *Non-Ferrous Metals Industry Unionism* (New York: Cornell University 1954). There was an ongoing struggle within Mine Mill between the communists and the anti-communists.

28 *Ibid.*, chapters 18 and 19. In the United States, after Mine Mill was expelled from the CIO in 1949, its jurisdiction was awarded to the USWA and the UAW.

29 Reid Robinson, 13 February 1942, letter to Philip Murray, file 2900, J.L. Cohen Papers, PAC

30 J.L. Cohen, 17 February 1942, letter to Reid Robinson, *ibid.*

31 Reid Robinson, 13 February 1942, letter to Philip Murray, *ibid.*

32 Larry Sefton, recording secretary, Local 240, personal interview with the author at Toronto, 20 June 1972

33 Transcript of taped interview with Eamon Park, assistant national director, USWA, p. 12

34 *Steel Labor*, 27 February 1942

35 *Canadian Unionist* (March 1942), p. 249

36 Logan, *Trade Unions in Canada*, p. 545

37 S. Crysdale, *The Industrial Struggle and Protestant Ethics in Canada* (Toronto: Ryerson Press 1961), p. 61

38 CCL memorandum, 27 February 1942, p. 1, private papers of Mary Sefton. Among the grievances were:
   1 those mentioned in the text;
   2 no adequate consultation with the NLSC about policies affecting workers;
   3 interpretations of PC 7440 in the Peck Rolling Mills and McKinnon Industries disputes;

4 the refusal of the government to enforce PC 7440 on the railways without first insisting on an abrogation of the employees' rights before paying the bonus;

5 Brunning's appointment in the National Steel Car dispute with his anti-labour reputation, the encouragement of a company union there and the refusal of the government to enforce its labour policy;

6 June 1941 amendment to the IDI Act restricting choices of representatives to conciliation boards;

7 the lack of labour representation on the IDIC and its decisions in the Kirkland Lake and Canada Packers disputes;

8 PC 5830 facilitating the use of troops in labour disputes on the presumption that sabotage had occurred in the Arvida strike, and the failure to rescind this order-in-council when it was established that the allegation of sabotage was unfounded;

9 PC 7307;

10 wage freeze by PC 8253 restricting negotiations and freezing inequalities in wage rates;

11 the Kirkland Lake situation;

12 the ruling by justice minister that the IDIC could not investigate discrimination at Research Enterprises Ltd, because it was government owned;

13 the discouragement of union shop agreements by the director general of labour relations in the Department of Munitions and Supply;

14 the fixing of wage rates in the shipyards in contravention of collective bargaining and wage policies.

39 *Ibid.*, p. 2
40 *Ibid.*, p. 3
41 *CCL Convention Proceedings 1942*, p. 8
42 CCL memorandum, 27 February 1942, p. 4, private papers of Mary Sefton
43 William Lyon Mackenzie King (WLMK) typescript diary, 27 February 1942, vol. 90, King papers, PAC
44 *Ibid.*
45 Daniel Coates, 'Organized Labor and Politics in Canada: The Development of a National Labor Code' (PHD dissertation, Cornell University 1973), p. 105
46 *Ibid.*, p. 106
47 CCL convention proceedings 1942, stenographic report
48 *CCL Convention Proceedings 1942*, p. 8
49 CCL convention proceedings 1942, stenographic report, p. 321
50 *Ibid.*, pp. 324, 326, 327
51 *CCL Convention Proceedings 1942*, p. 38
52 CCL convention proceedings 1942, stenographic report, p. 452

53 *Ibid.*, p. 454
54 *Ibid.*, p. 457
55 *Ibid.*, p. 1
56 *Ibid.*, p. 21
57 *Ibid.*, p. 29
58 *Ibid.*, p. 42
59 Gerald Caplan, *The Dilemma of Canadian Socialism* (Toronto: McClelland and Stewart Ltd 1973), p. 95
60 CCL convention proceedings 1942, stenographic report, p. 44
61 *TLC Convention Proceedings, 1942*, p. 214
62 Coates, 'Organized Labor and Politics in Canada,' p. 105
63 This was Meighen's second time as leader of the Conservative Party.
64 *Toronto Star*, 10 February 1942
65 J.W. Pickersgill, *The Mackenzie King Record*, vol. I (Toronto: University of Toronto Press 1960), p. 348
66 David Lewis, 2 January 1942, letter to C.H. Millard, vol. 196, national Co-operative Commonwealth Federation (CCF) papers, PAC
67 C.H. Millard, 31 December 1941, letter to David Lewis, vol. 196, national CCF papers, PAC. 'Every trade unionist in the Ontario section with any official standing ... including the A.F. of L'ers, and both the International and National unionists with the Canadian Congress are agreed, and most of them insistent that Mitchell should be opposed.'
68 Provincial executive and council minutes, 5 December 1941 and 2 January 1942, Ontario CCF papers, Queen's University; Reginald Whitaker, *The Government Party* (Toronto: University of Toronto Press 1977), p. 135. For political reasons, the Liberal Party contributed financially to the CCF campaign in South York.
69 Provincial executive and council minutes, 5 December 1941, Ontario CCF papers, Queen's University
70 Provincial executive minutes, 13 February 1942, Ontario CCF papers, Queen's University. Mitchell won with 45.6 per cent of the vote. The CCF came third with 21.7 per cent of the vote. Its vote had tripled since 1940 when it received 7.73 per cent.
71 Gad Horowitz, *Canadian Labour in Politics* (Toronto: University of Toronto Press 1968), p. 70
72 Provincial executive minutes, 2 January 1942, Ontario CCF papers, Queen's University
73 *Toronto Star*, 5 February 1942
74 *Ibid.*, 10 February 1942
75 David Lewis, 2 September 1942, letter to Clarie Gillis, vol. 95, national CCF papers, PAC

76 Clarie Gillis, 11 September 1942, letter to David Lewis, *ibid.*
77 Caplan, *The Dilemma of Canadian Socialism*, p. 95
78 Coates, 'Organized Labor and Politics in Canada,' p. 101
79 *Ibid.*, pp. 105, 110
80 *Ibid.*
81 *Ibid.*, pp. 138–40, 220
82 Bob Carlin, financial secretary, Local 240, transcript of taped interview with the author at Gowganda, August 1973, p. 34
83 *Ibid.*, p. 35
84 Coates, 'Organized Labor and Politics in Canada,' pp. 138–40
85 'Summary of Activities of the Labour Court, June 14, 1943 to December 31, 1943,' Ontario Labour Court, Ontario Department of Labour, Provincial Archives of Ontario (PAO)
86 Coates, 'Organized Labor and Politics in Canada,' p. 219
87 Eugene Forsey, 'Mr. King and the Government's Labour Policy,' *Canadian Forum* (November 1941): 232
88 H.A. Logan, *Trade Unions in Canada* (Toronto: Macmillan Company of Canada Ltd 1948), p. 546
89 Coates, 'Organized Labor and Politics in Canada,' p. 225
90 CCL convention proceedings 1942, stenographic report, p. 454
91 *Canadian Unionist* (February 1942), p. 200

Bibliography

PRIMARY SOURCES

Public Archives of Canada
Canada, Department of Labour papers
Canadian Labour Congress (CLC) papers
J.L. Cohen papers
Co-operative Commonwealth Federation (CCF) papers
William Lyon Mackenzie King (WLMK) diary
William Lyon Mackenzie King papers

Province of Ontario Archives
Mitchell Hepburn papers
Ontario Department of Labour papers
Ontario Attorney General's papers

United Church of Canada Archives, Victoria College, University of Toronto
Board of Evangelism and Social Service papers
J.R. Mutchmor papers

McMaster University Library
J.N. 'Pat' Kelly papers
Labour pamphlets

Queen's University Library
Ontario Co-operative Commonwealth Federation papers

PRIVATE COLLECTIONS

Labour Council of Metropolitan Toronto, minutes of Toronto Trades and Labour Council meetings 1940–42
Larry Sefton papers. A sparse collection of documents in the author's possession.
Mary Sefton papers. A sparse collection of documents in the author's possession.
United Steelworkers' files and correspondence in the USWA national office in Toronto

GOVERNMENT PUBLICATIONS

Canada. Census 1931, 1941
– Department of Labour. *Labour Gazette* 1937–49 (published monthly)
– Department of Labour. *Labour Organizations in Canada* 1937–49 (published annually)
– Department of Labour. *Strikes and Lockouts in Canada* 1937–49, 1976–77
– Royal Commission to inquire into the events that occurred at Arvida, Quebec, 1941
Canadian Annual Review of Public Affairs 1937–38
Debates. House of Commons 1939–45 (on matters pertaining to labour relations)
International Labour Organization, E.J. Riches, *Labour Conditions in War Contracts* (Montreal 1942)
Ontario. Proceedings of Select Committee re Bargaining between Employers and Employees (1943)
– *Sessional Papers* 1941–42
Orders-in-council: PC 2685, PC 2686, PC 5922, PC 7440, PC 4020, PC 4844, PC 7307, PC 10802, PC 1003

INTERVIEWS

'Doc' Ames (Kirkland Lake, Ontario, 1973)
Bob Carlin (Gowganda, Ontario, 1973)
Kay Carlin (Gowganda, Ontario, 1973)
Fred Dowling (Toronto, Ontario, 1974)
Dick Hunter (Kirkland Lake, Ontario, 1973)
Tom McGuire (Livermore, California, 1976)
Eamon Park (London, England, 1974)
Arnold Peters, MP (Kirkland Lake, Ontario, 1973)
Larry Sefton (Toronto, Ontario, 1972)

NEWSPAPERS AND PERIODICALS

Canada. Department of Labour. Press clippings files 1940–43
Canadian Forum 1937–47
Canadian Mining and Metallurgical Bulletin 1939–45
Canadian Unionist 1939–45
Daily Clarion 1936–38
Financial Post 1939–42
The Globe and Mail 1937–47
Northern Miner 1939–42
Northern News (Kirkland Lake) 1939–42
Sudbury Star 1940–43
Steel Labor (Canadian edition) 1940–42
Timmins Press 1940–42
The Toronto Star 1941–42
United Church Observer 1940–42

THESES AND UNPUBLISHED MANUSCRIPTS

Alway, Richard M. 'Mitchell F. Hepburn and the Liberal Party in the Province of
    Ontario 1937–43.' MA thesis, University of Toronto 1965
Coates, Daniel. 'Organized Labor and Politics in Canada: The Development of a
    National Labor Code.' PHD thesis, Cornell University 1973
Copp, Terry. 'The Impact of Wages and Price Control on Workers in Montreal
    1939–47.' Unpublished paper, Wilfrid Laurier University, May 1976
Lang, John B. 'A Lion in a Den of Daniels – A History of the IUMMSW in
    Sudbury, Ontario 1942–62.' MA thesis, University of Guelph 1970
Martin, William Arnold. 'A Study of Legislation Designed to Foster Industrial
    Peace in the Common Law Jurisdiction of Canada.' PHD thesis, University of
    Toronto 1954
Pentland, H.C. 'A Study of the Changing Social Economic and Political Back-
    ground of the Canadian System of Industrial Relations.' Unpublished study for
    the Task Force on Labour Relations, Ottawa 1968

Index